Inventing Atmospheric Science

Inventing Atmospheric Science

Bjerknes, Rossby, Wexler, and the Foundations of Modern Meteorology

James Rodger Fleming

The MIT Press
Cambridge, Massachusetts
London, England

This book was set in Stone Sans and Stone Serif by Toppan Best-set Premedia Limited. Printed and bound in the United States of America.

Library of Congress Cataloging-in-Publication Data is available.

ISBN: 978-0-262-03394-7

10 9 8 7 6 5 4 3 2 1

Dedicated to my teachers of history and science

Technology is cutting the Gordian Knot instead of untying it.
—Carl-Gustaf Rossby, "Current Problems in Meteorology" (1956)

Contents

Acknowledgments

This project began during a Charles A. Lindbergh Fellowship year at the Smithsonian National Air and Space Museum. It continued with a sabbatical year granted by Colby College, during which I held a Roger Revelle Fellowship from the American Association for the Advancement of Science and was a visiting scholar at the National Academies of Science and the Woodrow Wilson International Center for Scholars. The first draft of the book was written during another sabbatical year granted by Colby College, during which I was a visiting scholar at Columbia University. The final manuscript went to press while I was a guest of the International Meteorological Institute in Stockholm. Throughout the entire process, Miyoko was at my side, providing constant love and support.

Thanks to a number of individuals who rendered invaluable services: Gunnar Ellingsen, Anders Persson, Nicole Sintetos, John Sweney, Magnus Vollset, and three anonymous referees read the manuscript and provided valuable comments; Abe Krieger and "Victoria" read it aloud; David DeVorkin sponsored my Smithsonian fellowship and shared his interests in rockets and Wexler; Gunnar Ellingsen provided Norwegian translations of Bjerkenes's letters; Manny Gimond scanned the oversize maps; Honoria Bjerknes Hamre granted access to the Bjerknes Family Collection; Winnie Lau and Diana Schneider provided inspiration, encouragement, and valuable suggestions; Roger Launius suggested themes and framing; John M. Lewis generously shared his personal research collection on Rossby; Lou McNally reanalyzed the Anne Louise Beck weather maps; Carolyn Morgan traced the genealogy of Beck; Anders Persson, Norman Phillips, and Sverker Sörlin provided original documents and sage advice; Alan Robock contributed historical insights and scientific expertise; H. Thomas Rossby and

Michael Tjernström shared their perspectives on Carl-Gustaf Rossby; and Magnus Vollset assisted with the Bjerknes family letters.

Robert Canning, Abe Krieger, Lauren Lacy, Erin Love, Eoin McCarron, Ashley Oliver, and Alexa Williams served as dedicated research assistants. Dan Barbiero and Janice Goldblum (U.S. National Academy of Sciences Archives); Nina Korbu (Nasjonalbiblioteket, Oslo), Amy Malaise (Santa Rosa Junior College Archives), Nora Murphy (MIT Institute Archives and Special Collections), Gunnar Ellingsen and Yngve Nedrebø (Statsarkivet i Bergen), Karl Grandin (Kungliga Vetenskapsakademien Archives), Yngve Nilsen (Bergen Geophysical Institute), and John Strom (Carnegie Institution for Science Archives) added personal touches to the archival work. I also acknowledge support from the archivists at the Institute for Advanced Study, U.S. Library of Congress Manuscripts Division, U.S. National Archives and Records Administration, the National Center for Atmospheric Research, Stockholm University, and the University of Chicago and support from the librarians in the print and electronic collections of Colby College Libraries, Columbia University Libraries, Hathi Trust, Linda Hall Library, National Oceanic and Atmospheric Administration (NOAA) Central Library, Smithsonian Institution Libraries, University of California–Berkeley Libraries, and the U.S. Library of Congress.

I presented seminars on this work at the American Meteorological Society, Carnegie Mellon University, Centre National de la Recherche Scientifique (CNRS) in Meudon-Bellevue, Colby College, Colorado State University, Columbia University, Community Collaborative Rain, Hail and Snow Network (CoCoRaHS), History of Science Society, Rachel Carson Center, Rotman Institute of Philosophy at the University of Western Ontario, Rutgers University, Sigma Xi at Allegheny College, Smithsonian National Air and Space Museum, Stockholm University, Swedish Royal Institute of Technology (KTH), University of Bergen, University of Illinois at Champagne-Urbana, University of Manchester, University of Pennsylvania, and University of South Carolina.

This book would not have been completed without the support and encouragement of Susan Wexler Schneider, who generously shared the Wexler family papers and her personal stories with me.

Finally, special acknowledgments to Margy Avery, Deborah Cantor-Adams, and Katie Helke, editors extraordinaire.

1 Introduction

"The goal of meteorology is to portray everything atmospheric, everywhere, always." This striking proclamation of 1960 by John Bellamy and Harry Wexler captured the excitement of the moment, three months after the successful launch of *TIROS 1*, the first weather satellite.[1] In a larger historical sense, meteorological researchers everywhere always held this sentiment. In the middle of the nineteenth century, the electrical telegraph opened up new possibilities for data collection and the distribution of reports and forecasts.[2] In the first decade of the twentieth century, researchers pinned their hopes on wireless telegraphy and the dawning of the aerial age. Several decades later, radio, aviation, rockets, digital computing, and Earth-orbiting satellites had opened up entirely new research horizons. Each generation of atmospheric researchers aspired to a more global meteorology, zealously incorporating capabilities provided by new technologies into their science as they worked to link theory with practice. Members of each generation experienced, in their own ways, the heady feeling that they were the direct beneficiaries of new technological breakthroughs and that they stood on the brink of a major revolution in the science and practice of meteorology. Their goal was to produce accurate information about the state of the entire atmosphere, complete mathematical portrayals of its varied and changing states, and useful and timely forecasts of its near- and long-term future. Their soaring aspirations faced a multiplicity of crushing limitations exacerbated by war, bureaucracy, economic downturns, prejudice, technological limitations, and, perhaps most important, the ultimate realization that perfectly accurate measurements and perfectly accurate forecasts would never be possible.

Woven throughout this book are sections that focus on the technologies that transformed meteorology—radio, aerospace, and nuclear. Radio's lineage involves wireless, continuous wave radio, radiosondes, radar, digital computers, and the telecommunications infrastructure of the space age. Aerospace, broadly interpreted to include the vertical dimension of exploration and transportation, encompasses instrumented kites and balloons, commercial and military aviation, sounding rockets, and weather satellites. Nuclear issues date to the discoveries of 1895–96, followed by several decades of scientific interest in ionizing radiation in the atmosphere. After 1945, atmospheric detonations of nuclear weapons injected radioactive tracer particles that circled the planet in the wind regimes and descended to Earth as fallout, providing evidence for other studies of atmospheric chemistry and aerial pollutants.

If most informed observers today were asked, "What technologies made meteorology modern?," they might respond, "Computers and satellites." There is a longer perspective, however—from Marconi wireless to telecommunications and digital computing, from the Wright brothers' flier to sounding rockets and weather satellites, and from Becquerel and roentgen rays to the radioactive tracers of the age of outdoor nuclear testing. In every generation, meteorologists appropriated these technologies, eagerly and almost immediately, in their ongoing quest for new observational and modeling capabilities.

This is a book about the future—the historical future as three interconnected generations of atmospheric researchers experienced it and envisioned it in the first part of the twentieth century. The main story line involves Vilhelm Bjerknes, Carl-Gustaf Rossby, Harry Wexler, and their associates who worked at the center of meteorological research from just before 1900 to just after 1960 (figure 1.1). Bjerknes mentored Rossby, and Rossby trained Wexler. Over those sixty years, the three men worked at the cutting edge of geophysical fluid dynamics, weather analysis and forecasting, and the appropriation of new technologies. The opportunities were great, but the scale and complexity of the tasks were daunting. Their lives and careers were intertwined, and their life stories, told here in sequence, provide a human element to the history of a complex, interdisciplinary field. Collectively, their portraits provide insights into the development of meteorology in the twentieth century. All three were outstanding scientists, institution builders, and extraordinary networkers. They were not quintessentially geniuses but representatives of a much larger meteorological enterprise that came to be known circa 1960 as atmospheric science.

Figure 1.1
Left to right: Vilhelm Bjerknes (1862–1951), Carl-Gustav Rossby (1898–1957), and Harry Wexler (1911–1962). Photos taken in the early 1930s.

The book is grounded in archival sources, many of them newly uncovered—such as Bjerknes's dealings with the Carnegie Institution, Rossby's scattered letters, and Wexler's official and family papers.[3] It sheds new light on the three men, their interactions with their colleagues, and the ways that meteorology was transformed into atmospheric science in an era spanning two world wars and the Cold War, the radio age, the early space age, and the atomic age. The book introduces new protagonists, such as Anne Louise Beck, who was the first person to attempt to bring Bergen school methods to the United States. It provides new perspectives on hydrodynamics, the theory and practice of short- and long-range forecasting, dynamic climatology, global environmental change, and the dawn of chaos theory.

This is big picture history. It uses the lives and careers of central protagonists to map the history of weather research and weather-related technologies and chart the pathways by which meteorology became atmospheric science. There is grandeur in this subject. After all, we are talking about the behavior of the atmosphere, both locally and globally—how scientists came to measure it with new tools and understand its complexities, transforming theory and practice in meteorology and in oceanography as well. As Robert P. Multauf once observed, "Meteorology is in many ways unique among the sciences, and its history would perhaps contribute more to our understanding of the place of science in human history than that of any other special science."[4]

Big picture history of science is of large import, wide expanse, and *longue durée*.[5] This book examines not all science and technology but atmospheric science and technology, not all nations but a set of career trajectories represented by the three protagonists and their close associates, and not all of modernity but the technologically innovative first six decades of the twentieth century. Presented here is a pretty big history, with foregrounded details accessible through the biographies of the central players and a sweeping synthesis with considerable heterogeneity. This is not a grand narrative of human experience or even of human aerial experience but a set of connected stories with theoretical coherence, central actors, breadth of vision, and unity of action. It is a story with warm-blooded protagonists—a story with a beginning, a middle, an end, and a new beginning. It provides opportunities to challenge received views and to discover new aspects of histories that have not yet been told. Big picture framing is a metaphor. Some of the biggest or most important pictures dominate their galleries, even their museums. This work is a skyscape of atmospheric science and technology, an invitation to think creatively about a unique interdisciplinary science, its prospects and limitations, and the contribution it makes to understanding the place of science in recent human history.

Meteorology was "on the move" in the 1960s with Wexler's involvement in the launch of the first weather satellites, the founding of the National Center for Atmospheric Research, and the planning of international programs for improved weather forecasts and for global atmospheric research.[6] Meteorology was on the move in the 1950s, as well, in the form of sounding rockets that collected high-altitude data and reached the top of the atmosphere to photograph storms; digital computers that produced experimental and operational numerical weather predictions; and an increasing focus on solar-terrestrial relationships, glaciology, and chemical meteorology institutionalized during the International Geophysical Year of 1957–58, an eighteen-month period in which sixty-seven nations conducted research at both poles, in outer space, and at many locations in between. Meteorology was on the move in the 1940s as weather radar came on line, as thousands of weather cadets were trained to meet the needs of aviation in a global war, as atmospheric nuclear detonations provided radioactive tracers for upper-level winds, as landmark programs in meteorological science emerged at the University of Chicago and Stockholms Högskola, and as the first issues of the new research journals *Tellus* and *Journal of Meteorology* appeared.

Meteorology was on the move in the 1930s, even during a worldwide financial depression, as radio scientists probed the upper layers of the ionosphere and hordes of radiosondes—miniaturized radio sensors attached to sounding balloons—beamed back measurements from their ascents into the lower stratosphere, providing Rossby with the information necessary to formulate his theory of planetary waves. In the 1920s, the Bergen school flourished, Lewis Fry Richardson published his numerical experiment, aviation weather forecasts and radio broadcasts proliferated, and the meteorology department at the Massachusetts Institute of Technology opened for graduate training.

Meteorology was on the move in the 1910s in support of total warfare in a number of ways—providing observations and forecasts for nascent aviation, calculating trajectories for artillery, and launching and avoiding poison gas attacks. By the end of the decade, the new Bergen school had identified the structure and family histories of cyclones moving along the polar front, and its practitioners had developed air mass analysis in service to aviation. In the first decade of the century, new communications possibilities opened up, heavier-than-air flight became a reality, balloon soundings reached and identified the stratosphere, and Vilhelm Bjerknes issued his famous manifesto calling for a program to treat weather forecasting as a problem in mechanics and physics.

Noi viviamo sommersi nel fondo d'un pelago d'aria: we live submerged at the bottom of an ocean of air.[7] This was true enough for Evangelista Torricelli in 1644 and for most of us since then, most of the time. But the twentieth-century transformations in transportation and communication changed our focus and our relationship to the atmosphere in fundamental ways. They changed how we travel, as innovations such as the airplane rendered new significance to visibility, winds, and weather far above the surface and far afield; and they changed how we send and receive information, via wireless, radio, and television—so much so that the twentieth century marked the dawn of a new electromagnetic social environment.

We live in an ocean of electromagnetic radiation, of which only a small portion is visible to us. Earth is bathed in the electromagnetic emissions of the sun and the galaxy, Earth's magnetic field creates brilliant auroral displays, and charge separation in the atmosphere generates powerful electrical storms. Life itself evolved in this environment. It was only at the start of the twentieth century, however, that humanity began its active participation in the electromagnetic environment by generating and receiving signals. Meteorologists immediately appropriated wireless technologies and radio to disseminate observations and forecasts, probe the ionosphere, and conduct atmospheric soundings using radiosondes, They reveled in the new information being revealed by radar echoes from distant storms. Television emerged as a new technology for communication and visual imaging, and vacuum tubes and transistors functioned as electronic switches and logic gates in calculating machines and digital computers. Where would aviation and the space age be without the telecommunications capabilities provided by radio?

Most of the time, we live, work, and play at the bottom of the ocean of air, but twentieth-century humanity has ventured into it and even beyond it for recreational, commercial, scientific, and military purposes. New aviation and aero-space technologies fundamentally transformed the capabilities, aspirations, practices, and theories of atmospheric researchers over the course of six decades. Near the top of the ocean of air—far above where people live, work, and play—research aircraft collected routine data and flew into severe storms, miniaturized radio transmitters on sounding balloons soared into the stratosphere, radio waves bounced off the ionosphere, research rockets carried instruments to the edge of space, and weather satellites photographed clouds and scanned upwelling radiation for recognizable patterns, crossing national boundaries as they conducted their meteorological surveillance.

At the turn of the twentieth century, during the "first" atomic age, physicists wondered if the ionizing emissions from elements like radium and thorium were powering the global electrical circuit and possibly influencing weather systems. For a decade and a half, during the "second" atomic age of outdoor nuclear testing, meteorologists acquired new responsibilities to track nuclear debris worldwide, to monitor nuclear fallout, and to respond to public concerns that nuclear detonations were changing the weather. They also developed new tools and theoretical insights to address problems of local, regional, and global air pollution and atmospheric chemistry in general.

Maps and paths constitute shorthand ways of talking about methods. In the time-honored tradition of constructing maps of conditions on the surface or at five kilometers aloft, meteorologists were seeking a bird's eye, or synoptic, view of the weather that might reveal recognizable patterns— such as Helmholtz waves (instabilities between air masses along a mountain ridge), Bjerknes waves (wave cyclones propagating as families along the polar front), planetary waves (hemispheric-scale phenomena at high altitudes identified by Rossby that shape jet streams and interact with weather systems below), and global interactions and teleconnections between hemispheres (an agenda that motivated Wexler during the International Geophysical Year and the early space age).

The lives of the three major protagonists span a full century—from the birth of Bjerknes in 1862 to the death of Wexler in 1962. In 1898, in Stockholm, the year and location of Rossby's birth, thirty-six-year-old Vilhelm Bjerknes articulated his fundamental circulation theorem and began a correspondence with U.S. meteorologist Cleveland Abbe concerning the mechanics of Earth's atmosphere. Encouraged and inspired by his Swedish colleagues Svante Arrhenius and Nils Ekholm, Bjerknes began to formulate his approach to the basic problem of weather forecasting. Theories of meteorological dynamics had been under development for a number of decades, including Heinrich Wilhelm Dove's theory of the struggle between equatorial and polar currents, Herman von Helmholtz's work on the formation of a surface of discontinuity between cold polar air and warm equatorial air, and the asymmetric cyclone model of Henrik Mohn, which he used as the basis for storm warnings and occasionally weather forecasts.[8]

At the turn of the twentieth century, meteorologists focused largely on surface weather conditions; they had a sparse number of stations connected by telegraph lines, no observations over the oceans or unpopulated areas, and few upper-air measurements. Motorized flight was still several years in the future, and although instrumented kites had reached a record altitude of 4,850 meters and a piloted balloon had traveled to 10.3 kilometers, the Eiffel Tower was still considered one of the best upper-air measurement platforms available. Wireless telegraphy was just emerging, rockets were fireworks, Earth had one satellite known as the moon, and computers were people.

In 1911, the year of Wexler's birth, aviator Cal Rogers completed the first transcontinental airplane flight—from New York City to Pasadena, California—in eighteen days in a Wright brothers' biplane. He was supplied with handcrafted forecasts from a U.S. Weather Bureau firmly rooted in nineteenth-century ideas and practices. In the same year, Bjerknes and his collaborators published the landmark second volume of their magnum opus, *Dynamic Meteorology and Hydrography*, under the sponsorship of the Carnegie Institution of Washington. Meteorology was entering a period of rapid change on theoretical, practical, and technological fronts.

The stories are presented here, wherever possible, in the voices of the protagonists—three lives in research woven into the fabric of the twentieth century. The narrative reveals their opportunities and challenges; their trips, transitions, and career trajectories; and the contributions of the three protagonists and their many colleagues. These scientists developed new technical concepts involving the fundamental laws of hydrodynamics and thermodynamics. The concepts are used here sparingly without formal mathematics and are explained in common language in ways meant to buoy and clarify and not bog down the narrative. This book focuses on the history of atmospheric research and research practices, not on the daily grind of Weather Bureau officials and forecasters trying to stay one step ahead of the weather. However, it is duly recognized that the interplay of theory and practice was important and synergestic in this period.

Chapter 2 focuses on the perspectives, associates, roots, and legacy of Vilhelm Frimann Koren Bjerknes (1862–1951), a theoretical physicist nurtured by his father, trained by Jules Henri Poincaré and Heinrich Hertz, and inspired by the hydrodynamic theorems of Helmholtz and Lord Kelvin. The meteorological story line begins in Stockholm in the 1890s, where Bjerknes began to apply James Clerk Maxwell's equations for electricity and magnetism to dynamic meteorology and hydrodynamics. Confronted by the dearth of reliable observations and the absence of any unifying theory, Bjerknes wanted to put meteorology on solid observational and theoretical foundations. In 1904, he wrote that the necessary and sufficient conditions for a rational solution of the problem of meteorological prediction include a sufficiently accurate knowledge of the state of the atmosphere at a certain time and a sufficiently accurate knowledge of the laws according to which one state of the atmosphere develops from another. These statements bring together massive programs in observation, theory, and forecasting and are really philosophical statements of faith in rational mechanics, raising the fundamental question of how far we can see into the future. Starting from a detailed, if not perfect, set of atmospheric measurements, it should be possible to take a finite, if not perfect, step forward using the time-dependent equations of atmospheric motion. This is the Gordian knot of meteorology, an intertwined tangle of nonlinear influences that, if untied, would provide prevision of the weather for ten days, of seasonal conditions for next year, and of climatic conditions for a decade, a century, a millennium, or longer. Bjerknes's use of the word *sufficiently* in these statements tempered his determinism. The term undoubtedly derives from his exposure to the lectures of Poincaré and to his personal experiences in trying to measure, with any precision, the initial state of the atmosphere over an extended area and trying to predict, with any accuracy, the future state of the weather. His neo-Laplacian program set the agenda for the next six decades and beyond.

The main story line follows Bjerknes's career to Stockholm (1893), Kristiania (1907), Leipzig (1913), Bergen (1917), and back to Kristiania, now Oslo (1926). It documents his important visits to Paris, New York, Washington, DC, London, and Pasadena and his trajectory from thinking about ideal fluids in relation to electromagnetism to working with the real fluid motions of the atmosphere and ocean.

Under the leadership of Bjerknes in Bergen, his son Jacob, Halvor Solberg, Tor Bergeron, and many others established the polar front theory of midlatitude cyclones and the practice of air mass analysis. They worked diligently to spread their theoretical insights internationally, to collect and interpret more and better observations at higher and higher altitudes, and to codify graphical techniques of air mass analysis and weather forecasting for the use of national weather services. Following a graduate fellowship year studying with Bjerknes, twenty-five-year-old Anne Louise Beck published her thesis work in the *Monthly Weather Review*, becoming the first person to bring Bergen methods to the attention of the U.S. Weather Bureau in a systematic way. Collectively, Bjerknes and his associates championed something larger than themselves, something useful and entirely new.

The story continues in chapter 3 with Carl-Gustaf Arvid Rossby (1898–1957), arguably the most influential and innovative meteorologist of the twentieth century. Rossby was a gregarious Swede, a free spirit, a mover and a shaker, a builder of both institutions and theories, the center of geophysical attention wherever he went. He spent two years with Bjerknes but was very much his own man. He brought Bergen school methods of air mass analysis to the United States, established the first commercial airline weather forecasting system, built the first graduate program in meteorology at the Massachusetts Institute of Technology in 1928, and established formal relations with oceanographers at Woods Hole in 1930. He supervised the training of literally thousands of aviation cadets during World War II, developed a general theory of long waves and jet streams in the upper atmosphere, and established the University of Chicago's department of meteorology and the University of Puerto Rico's Institute of Tropical Meteorology. After the war, he enhanced the American Meteorological Society's professional standing by founding the *Journal of Meteorology* and, working closely with Jule Charney, stimulated work on numerical weather prediction at the Institute for Advanced Study. In the final decade of his life, Rossby returned to his native Sweden, where he established the International Meteorological Institute in Stockholm. There he supported the first operational numerical forecast in the world, generated by the BESK computer, and moved into issues of geophysics and global pollution by founding the international environmental journal *Tellus*. He was a dynamo, always the focus of a moving meteorological and oceanographic seminar. Chester Newton, a former student at Chicago, recalled, "When Rossby was in town, the department was in tumult. There were interesting people coming by, there were things happening every day, and—it was exciting but exhausting."[9]

Rossby had many notable students, none of them more energetic, more accomplished, or better documented than Harry Wexler (1911–1962), the focus of chapter 4. Wexler, the son of Russian Jewish immigrants, majored in mathematics at Harvard College, studied meteorology at MIT, and joined the U.S. Weather Bureau at age twenty-three. His early career experiences paralleled the development of meteorology: he worked to bring air mass analysis to the United States, applied new operational forecasting techniques at airport stations, and, as an officer in the U.S. Army Air Forces, trained a generation of weather cadets during World War II. In 1944, he became the first scientist to fly into a hurricane. After the war, the thirty-four-year-old Wexler rejoined the U.S. Weather Bureau as head of research, nurturing every new technological development and every new international program of relevance to the atmosphere. In this capacity, he worked to incorporate radar, rockets, nuclear tracers, digital computers, and satellites into meteorological practice. He demonstrated scientific leadership as a member of (and in many cases, chair of) many national and international geophysical advisory committees. Wexler's vision was global, as evidenced by his leadership of a massive interdisciplinary research program as chief scientist for U.S. Antarctic programs during the International Geophysical Year. He conducted experiments and coordinated measurements on the Antarctic ice sheet and established the Weather Bureau's Mauna Loa Observatory, where he supported the first measurements of carbon dioxide levels. In 1962, the year of his untimely death at age fifty-one, weather satellites were orbiting the earth every ninety minutes, digital computers were cranking out daily forecasts, and both superpowers were detonating hydrogen bombs in space. Wexler had been to Geneva three times on assignment from the Kennedy administration, negotiating the World Weather Watch, a lasting international agreement on the peaceful exchange of weather information. Wexler was on top of his science, a leader in new techniques and technologies, and a respected international figure.

At a meeting of the National Academy of Sciences Committee on Meteorology in 1956, originally established to advise the Weather Bureau, Rossby suggested that the definition of *meteorology* be enlarged to include the role of the atmosphere as a milieu and as an environment. He called for scientists from other fields to work together on specific problems and to take advantage of new techniques, new challenges, and new opportunities at hand and on the immediate horizon. Within two years, under the administrative leadership of physicist and engineer Lloyd Berkner, membership in the renamed Committee on Atmospheric Sciences expanded its core membership of meteorologists to include mathematicians, chemists, atomic physicists, and space scientists. Since that time, *atmospheric science* has become an umbrella term covering the explosion of research specialties in meteorology and climatology, cloud physics, atmospheric chemistry, and the dynamics of Earth's atmosphere from the surface to the edge of space and the surface of the sun. A new interdisciplinary field was emerging, designed and run by committee, motivated by new opportunities and heightened international tensions, the recipient of massive new federal funding. This is the focus of chapter 5, which chronicles how the University Corporation for Atmospheric Research, the National Center for Atmospheric Research, and international programs in weather service and weather research got their start.

In 1960, Edward N. Lorenz identified a fundamental chaotic limit to our ability to forecast the future. It was a game changer, a brick wall blocking prevision. Lorenz noted that the nonlinear nature of the equations used in numerical weather prediction would lead to multiple solutions from essentially identical initial states. Small perturbations caused large departures. By *chaos*, Lorenz meant something precise: the behavior and the final state of a dynamical system described by a set of nonlinear equations are sensitive to its initial state. In chaos theory, future states of the weather and climate then become identifiable with the attractor of the dynamical system—but the dynamical system may have more than one attractor! No matter how sophisticated the technology becomes, chaos theory held that perfectly accurate measurements and perfectly accurate forecasts would never be possible. In computer weather modeling, this meant there is no way to make accurate long-term predictions.

Threads from the lives and work of Bjerknes, Rossby, Wexler, and their associates interlink all of the chapters. The bulk of their work—founding new institutions, generating new ideas, recruiting new students, using new technologies, fulfilling new needs, and taking new observations—helped make meteorology global. They were highly motivated individuals working to advance their careers and reputations; they were also public-minded scientific entrepreneurs or, more precisely, *entreprenours,* an obsolete Middle English term describing a person who undertakes, manages, controls, champions, and basically calls into existence something new under the sun, typically through an effort involving the marshaling of people, ideas, and resources. Many of their close associates were team players too, especially those who led and staffed the committees that deliberated and formulated the institutions of the atmospheric sciences after 1957.

There are relatively few twentieth-century histories of atmospheric research and its supporting technologies. Those that exist are quite strong—admirable, fundamental, standard works focusing on particular eras or technologies. Some histories, however, are narrowly focused. Many emphasize institutions, bureaucracies, and weather services rather than weather research. Of the historical articles written by scientists, most provide reliable, if sometimes fragmentary and often dauntingly technical glimpses into particular issues.[10] This book is unique, offering a big picture history of atmospheric research and technology in the first six decades of the twentieth century. It provides intimate and intertwined details of the lives and work of three major protagonists over three generations interwoven into a larger tapestry of soaring aspirations and crushing limitations.

2 Bjerknes

In Transit, Paris 1881

Tout chemin mène à Paris et tout chemin part de là. C'est la Rome des temps nouveaux.
—Ximénès Doudan[1]

In the summer of 1881, nineteen-year-old Vilhelm Bjerknes and his father, Professor Carl Anton Bjerknes (1825–1903), traveled to Paris on business.[2] The elder Bjerknes had received an invitation from the noted physicist Eleuthère Mascart to display his hydrodynamical instruments at the International Electrical Exposition. Father and son left Kristiania on July 15, bound for Le Havre, on the Norwegian steamer *SS Kong Magnus*. They arrived in Paris by train, but they arrived too soon. Several days after the scheduled opening, the exhibition hall was still "a big mess," with workers scrambling to arrange everything and the Bjerkneses trying to stretch their limited travel funds.

Finally, the new age of electricity opened for public display in the cavernous Palais de l'Industrie along the Champs Élysées on a site about halfway between the Louvre and the Arc de Triomphe. Fairgoers could book a ride on the Siemens electric railway from the Place de la Concorde to the exhibit hall, where they beheld the huge steam-powered dynamos that ran the machinery and a vast array of novel electric devices on display— batteries, boats, detonators, telegraphs, and novelties tended to by their inventors and promoters. Evenings under the glass and steel arch of the great hall were spectacular. The younger Bjerknes recalled "the whole palace shining in the most wonderful electric light," illuminated by the thin carbon wires that glowed inside the light bulbs of Hiram Maxim and Thomas Alva Edison and swept by the beams of two massive searchlights.

When Vilhelm grew tired of the overwhelming dazzle, he retired to a peaceful corner of the building where the history of electrical research was on display—the manuscripts of Luigi Galvani; the homemade original instruments of Alessandro Volta, Hans Oersted, Andrè-Marie Ampère, and Michael Faraday; and the clear didactic message that basic scientific research in one generation would result in surprising technological benefits in the next.

At the International Electrical Exposition, there were hordes of people to impress, deals to be made, delicacies to be sampled, pockets to pick, and fortunes at stake. The daily newspapers were there in force, as were the technical press and a host of other media. *Scientific American* commented: "Certainly nothing has ever been done before so effectually to popularize science, and to render the masses familiar with the effect, however ignorant they may be of the cause, of this marvelous invention."[3] One of the most popular attractions at the exhibition was the demonstration of the newly improved Adler telephone, or "speaking telegraph." Each evening, a bank of eighty telephones transmitted the singing on the stage and the music in the orchestra of the Grand Opera at Paris directly to a suite of four listening rooms at the Palais de l'Industrie. The masses stood patiently in long queues for hours, waiting their turns "alone together" at the telephone receivers.

The First International Congress of Electricians opened in September in conjunction with the exposition. It attracted some 250 delegates from twenty-eight different nations to debate and define standard units—the ohm, the volt, the ampere, the coulomb—and to showcase the new field of electrical engineering, which had recently emerged, "a lusty child of science and machinery."[4] Here the importance of standard units in science and engineering made a big impression on the youthful Bjerknes. The discussions at the conference were all electrical. Vilhelm Bjerknes did not turn his attention to things meteorological until the century was almost finished. In Paris, however, he witnessed the fundamentally transformative powers that technology exercised over daily life and over the practice of science.

In August, the elder Bjerknes had to return home on urgent business, leaving his son in Paris to staff the small Norwegian exhibit of laboratory models, which demonstrated fields of force and affinities between hydrodynamics and electrodynamics.[5] Carl Anton Bjerknes's goal was to develop a "hydrodynamic picture of the world" that explained the behavior of classical fluids by analogy to James Clerk Maxwell's electromagnetic field theory. He first formulated this idea as a little boy from Leonhard Euler's "Letters to a German Princess," where he was impressed by arguments opposing the doctrine of action at a distance and favoring the agency of the ether, the hypothesized subtle medium filling all of space responsible for transmitting wave motion. Euler wrote, "It appears in truth abundantly certain, that light is with respect to ether, what sound is with respect to air; and that the rays of light are nothing else but the shakings or vibrations transmitted by the ether, as sound consists in the shakings or vibrations transmitted by the air."[6]

Carl Anton was reminded of Euler's ideas many years later, in 1856, when he attended lectures on hydrodynamics by the German mathematician Gustav Lejeune in Göttingen. The question arose: "If two bodies are moving in a fluid, will not they, through the liquid as an intermediary, mutually act on each other's movements? And would not one observer who sees the bodies, but not the liquid, believe that he has seen an effect at a distance between the separated cells?" These questions brought him to set up the mathematical problem of the simultaneous motion of a system of spherical and cylindrical bodies in a homogeneous incompressible fluid. Because the elder Bjerknes lacked some of the needed mathematical skills to develop a complete theory, he devised laboratory demonstrations, while his son Vilhelm built and demonstrated them.[7]

The illustration of forces in fluids was a big hit with both elites and the general public. Young Bjerknes wrote at the time, "I can hardly dry the apparatus with a cloth before people come back to see more." The who's who of visitors to the table included David Hughes, inventor of the carbon microphone; Gaston Plante, of lead-acid battery fame; Hiram Maxim and Thomas Edison, coinventors of the light bulb; Alexander Graham Bell, inventor of the telephone; and engineers and industrialists Werner and William von Siemens. The nineteen-year-old Bjerknes lectured and demonstrated the apparatus to them and to noted scientists John Tyndall, Lord Rayleigh, A. A. Michelson, Gustav Kirchhoff, William Crookes, and Ernst Mach, among others. Hermann von Helmholtz and William Thomson (later Lord Kelvin), who would greatly influence Vilhelm's thinking about fluid mechanics, also visited the table. One rather mysterious and dignified visitor stayed an hour with his entourage, later sharing his card, which read "Marquess of Salisbury."

According to Vilhelm Bjerknes, everyone referred to the hydrodynamic display as "the best of the whole exhibition." George Forbes reported favorably on the demonstrations in the British journal *Nature*:

From a scientific and purely theoretical point of view there is no object in the whole of the Electrical Exhibition at Paris of greater interest than the remarkable collection of apparatus exhibited by Dr. C. A. Bjerknes of Christiania, and intended to show the fundamental phenomena of electricity and magnetism by the analogous ones of hydrodynamics. I will try to give a clear account of these experiments and the apparatus employed; but no description can convey any idea of the wonderful beauty of the actual experiments, whilst the mechanism itself is also of most exquisite construction. Every result which is thus shown by experiment had been previously predicted by Prof. Bjerknes as the result of his mathematical investigations.[8]

Carl Anton Bjerknes was awarded one of the eleven coveted *diplômes d'honneur* from the exhibition. The following year, Vilhelm Bjerknes published his first scientific paper in the Norwegian journal *Naturen*, a report of "recent hydrodynamic studies" and a description of the apparatus displayed in Paris.[9]

A Career in Flux

The relationship of Carl Anton and Vilhelm Bjerknes was both intimate and rather vexed. Scholars have portrayed the elder Bjerknes's career as stalled and Vilhelm's devotion to his father as somewhat pathological, emphasizing the son's need to distance himself from familial influences and set out on his own. The standard account blames the father for clinging to his son and almost stifling him, using Vilhelm to compensate for his own shortcomings and practically forcing Vilhelm to leave Norway. Another interpretation points to the demise of the electromagnetic ether theory and its effect on Vilhelm's hopes for a career in physics. There is much to commend such views.[10] Vilhelm wrote in 1944 that "the thread that my father began produced branches leading to meteorology, geophysics, and astrophysics," with meteorology dominating in Vilhelm's career due to contingent and practical necessities. This does not represent the musings of a son who had to break away from his father but rather a new view of force fields and the relationship between the electrodynamic and hydrodynamic equations and a move from a defunct ether theory to a much more complex but realistic analysis of the fields of force generated in the real fluids that constitute the atmosphere and oceans.[11]

Dynamic meteorology *is* applied physics, and there are cases to be made for continuity between the early hydrodynamical demonstrations and Vilhelm Bjerknes's mathematical development of geophysical hydrodynamics two decades later and also for a lifetime of intellectual continuity between father and son. After completing his master's degree with distinction in 1888, Vilhelm won an advanced scholarship, allowing him to attend Henri Poincaré's mathematical lectures in Paris. At the time, Poincaré was writing his three-volume *Les méthodes nouvelles de la mécanique céleste*, in which he emphasized the impossibility of attaining exact measurements, the insolvability of the three-body problem, and the use of fluid flow analogies to characterize all motions of mechanical systems. Bjerknes then worked for two years in Bonn with Heinrich Hertz on problems of electricity and magnetism.[12] Hertz's experiments involved transmission and reception of electromagnetic pulses and demonstrated wave effects such as reflection, refraction, polarization, and interference. Bjerknes held Hertz in high esteem. He admired the graphical method Hertz had invented for use in meteorology, which was a thermodynamic diagram depicting dry and saturated adiabats and lines of constant saturation on logarithmic scales of pressure and temperature. He also shared his Hertz's view on the importance of deterministic prevision in science, as codified in *The Principles of Mechanics*:

The most direct, and in a sense the most important, problem which our conscious knowledge of nature should enable us to solve is the anticipation of future events, so that we may arrange our present affairs in accordance with such anticipation.[13]

Hertz died at age thirty-six, leaving behind a wife and two young daughters, ages six and three, the younger having been born during Bjerknes's apprenticeship. Bjerknes remained in contact with the family over the years and, decades later, helped the Hertzes settle in England during times of political turmoil in Germany. Bjerknes returned to Kristiania in 1892 and completed his dissertation, "On the Movement of Electricity in Hertz's Primary Conductor."[14] The following year, he married Sophie Honoria Bonnevie, daughter of a prominent Norwegian civil servant, the first female student in mathematics at the university, and sister of the first female professor in Norway, Kristine Bonnevie. The couple settled in Stockholm, but Honoria was never able to pursue her interest in numbers and geometry other than through embroidery: "From then on, she was always waiting; waiting for letters, waiting for her husband to return, waiting for her sons to visit."[15] The Bjerkneses had four sons; the second, Jakob Aall Bonnevie (Jacob or Jack), worked in meteorology with his father.

In 1893 Vilhelm accepted a position as lecturer in mechanics at Stockholms Högskola and continued work on electrical resonance until 1895. He studied the damping of Hertzian waves and phenomena of electric resonance, "a notable contribution to radio development at a time when electric waves were in their infancy."[16] Ultimately, Bjerknes left the field of electromagnetism behind to study geophysics. Some have speculated he did so because of the death of Hertz, because the ether theory of electromagnetic propagation was on the ropes, or perhaps because the mainstream of physics turned to other discoveries such as cathode rays, roentgen rays, and Becquerel rays. Vilhelm Bjerknes's work with this father on hydrodynamics is typically seen as "an interesting blind alley," an assessment that has pretty much stuck.[17] There is another compelling interpretation, however, regarding his turn to meteorology and oceanography that originates with his associations in Stockholm in the 1890s.

Ideal Fluids and Real Fluids

Hermann von Helmholtz likened the behavior of the atmosphere to a number of flywheels, or rings of air—vortices or whorls—spinning around the earth's axis at different latitudes. The energy of the flywheels derives ultimately from unequal solar heating between low and high latitudes, which drives a slow meridional circulation from equator to pole. The flows are deflected by the rotation of the earth. Friction, in the form of turbulence at Earth's surface and in the free atmosphere, slows the flywheel circulation and maintains it at an average constant speed. Helmholtz was aware that waves formed along boundaries between air of different temperatures and densities. He personally observed and explained waves a few kilometers long that form along mountain ridges, but due to limitations in data collection and mapping, he could only theorize—albeit creatively and influentially—about large-scale and global surfaces of discontinuity in the free atmosphere. In 1858, Helmholtz published a paper on vortex motion (*Wirbelbewegung*) in an ideal fluid, assumed to be homogeneous and incompressible, in which the surfaces of constant pressure coincide with those of constant density. He demonstrated mathematically that in ideal fluids, vortices are neither created nor destroyed and they conserve their rotation (or lack thereof).[18]

Sir William Thompson (Lord Kelvin) hypothesized that matter is composed of vortex atoms in a perfectly homogeneous and subtle fluid, referred to as ether. He advanced a theory of the dynamical interaction of these atoms under the influence of forces analogous to those produced by electromagnetic circuits. The vibrations, rotations, deformations, and interactions of the vortices were sufficiently complex to account for the known chemical interactions of molecules as well as characteristics of their spectra. There was one huge problem, however: a series of negative observational experiments aimed at detecting the ether led physicists to abandon it in the first decade of the twentieth century. Physicist Frank Wilczek has recently called Kelvin's theory of vortex atoms in an ethereal fluid a "beautiful loser," sterile because it was replaced by quantum field theory.[19] Meteorology pursued a different path involving vortex theories of real fluids.

Bjerknes once confided that he had been engaged in hydrodynamic work for almost two decades "without feeling any affinity to the hopelessly capricious phenomena of meteorology." He had been blocked by the celebrated theorems of Helmholtz and Kelvin that held that vortices are eternal—that they never rise and never die away. Yet according to daily meteorological experience, atmospheric vortices incessantly arise and die away. After Bjerknes moved beyond the boundaries of classical hydrodynamics—from ideal fluids closely associated with the notion of an electromagnetic ether to real fluids (oceans and atmospheres)—he saw the way forward to treat surfaces of discontinuity and fluid vortices in all their glorious and turbulent complexity. His basic conceptual shift consisted of replacing his father's solid geometrical objects with "fluid bodies." He could then begin a systematic investigation of the formation of eddies around them and the propagation of disturbances in hydrodynamic fields of action on all scales.[20]

In Stockholm, Bjerknes's circle was composed of influential scientists interested in cosmic physics and oceanography. They included Otto Pettersson, chemist, oceanographer, rector of the university from 1893–1896, and co-convener of the first international hydrographic conference in Stockholm in 1899; Svante Arrhenius, founder and first secretary of the Stockholm Physical Society and rector of the University from 1897–1902; Nils Ekholm of the Swedish Meteorological Institute; and mathematician

Ivar Bendixson, whose work on curves defined by differential equations followed that of Siméon Denis Poisson. Bjerknes was an active member of the Physical Society and said that as the result of conversations with his "scientific friends," he happened to see that the theorems he had developed for an abstract fluid system had immediate bearing on the dynamics of the ocean and the atmosphere.[21]

In 1898, Bjerknes published his famous circulation theorem: "On a fundamental hydrodynamic theorem and its application especially to the mechanics of the atmosphere and the world ocean."[22] Here he pointed out that the assumptions of the original Helmholtz-Kelvin theorems—that the fluid is ideal and the flow is frictionless—were "far from the truth" and had only a limited application in meteorology. To attain more general theorems, he proposed starting with the equations of motion for frictionless fluids with no limiting assumptions about the fluid's density. In the ocean, the temperature and the salinity play the same roles in changing the density as do the temperature and the moisture in the atmosphere. He employed the Archimedean principle that differences in density and pressure generate motion and the theorem of Sir George Gabriel Stokes regarding divergence and curl in fluid flow. He then formally presented the mathematics of his circulation theorem and a geometrical method of visualizing the dynamic condition of a fluid. Using the circulation theorem, Bjerknes could also explain land and sea breezes, mountain and valley winds, and, qualitatively, cyclones and anticyclones as boundary phenomena propagating between the cold and warm winds of the general circulation.[23]

Bjerknes lectured on the circulation theorem at the Deutsche Naturforschergesellschaft in Munich in August 1899 and, at the request of Cleveland Abbe, chief scientist with the U.S. Weather Bureau, communicated his results to the *Monthly Weather Review*. He concluded that the theory was applicable to the atmosphere and the ocean treated together as one fluid medium and included an acknowledgment to his colleague Nils Ekholm for his assistance and inspiration in understanding the mathematical laws of vortex motion.[24] Weather Bureau theoretician Frank Bigelow commented that Bjerknes's line integrals in the atmosphere could be depicted on the well-known thermodynamic diagram of Hertz.[25] Such a connection to Bjerknes's departed mentor implies that they worked together on more than just electrodynamics when in Bonn.

The circulation theorem marked a turning point in Bjerknes's life as geophysical hydrodynamics and meteorology replaced electromagnetism. Bjerknes surrounded himself with assistants who based their research programs on developing the implications of the circulation theorem and pursued common interests in the dynamics of oceans and atmospheres. He and his colleagues shared "a tradition for the exchange of researchers and ideas" within the larger cooperative Nordic community of science.[26] The noted oceanographer and polar explorer Fridtjof Nansen and his distinguished associate Bjørn Helland-Hansen turned to Bjerknes to develop applications of the circulation theorem. Bjerknes and his talented students Johan W. Sandström and Vagn Walfrid Ekman conducted theoretical and observational studies of discontinuous internal surfaces in the oceans and the atmosphere and examined the wavelike disturbances that propagate along them. Bjerknes hypothesized that the mysterious phenomena of dead water, observed by Nansen, could be due to a wave internal boundary. When Ekman heard about the clockwise rotation of wind-driven currents one evening at a dinner party, "that same evening," he identified the fundamental dynamic law of current distribution that bears his name and later published the formal derivation using Bjerknes's circulation theorem. When Helland-Hansen came to Stockholm to study with Bjerknes, Sandström, was instrumental in applying the circulation theorem to the calculation of ocean currents.[27]

Bjerknes looked to the future, to the distant possibility of calculating the movements of the atmosphere and the ocean, to the more proximate hope of discovering rational dynamic principles that would allow for practical, graphical solutions and forecasts, and to the immediate need of obtaining a sufficient number of systematic observations from the oceans and from the upper strata of air with such regularity that they can be utilized in daily predictions. These projects were played out in Bjerknes's career in the years that followed.

The Carnegie Institution

By one estimate, the total endowment for science in the United States was $3 million in 1902, mostly concentrated at Harvard University, the University of Chicago, and the Smithsonian Institution.[28] In the same year, Andrew Carnegie provided an initial $10 million endowment for a new institution to alleviate what he termed America's "national poverty in science" and "to secure if possible for the United States of America leadership in the domain of discovery and the utilization of new forces for the benefit of man."[29] The Carnegie Institution of Washington listed among its goals cooperation with governments, universities, colleges, technical schools, learned societies, and individuals, "in any department of science, literature, or art." It aimed "to discover the exceptional man in every department of study whenever and wherever found, inside or outside of schools, and enable him to make the work for which he seems specially designed his life work."[30]

Cleveland Abbe, serving as chair of the Carnegie Advisory Committee on Meteorology, surveyed the long history of weather observations, maps, and charts. He predicted that the most important discoveries in the future would come not from the hundreds of experimentalists and thousands of observers or from business as usual but from "the pre-eminent right man" trained in mathematics and analytical mechanics who could elucidate the complex phenomena of the atmosphere and oceans: "Meteorology has attained a status analogous to that of astronomy in the century between Newton and Laplace. It is ready to receive a new leader and is looking for him."[31] Abbe was most enthusiastic about mathematical work on the dynamics of the earth's atmosphere in the tradition established by Helmholtz and cited, by name, Vilhelm Bjerknes and Nils Ekholm as scientists worthy of Carnegie support and, as a likely administrative leader, Robert S. Woodward, professor of mechanics and mathematical physics and dean of the faculty of pure science at Columbia University. Others also pointed out Woodward's leadership qualities, and he was appointed first president of the Carnegie Institution in 1904.

Not everyone providing advice agreed. Many well-known meteorologists recommended expanding existing observations and supporting laboratory work. Writing from Vienna, Julius Hann was of the opinion that the best an institution like Carnegie could do is support "works for which the methods are, in general, already established." He recommended the collection of world climatic data, especially of the intensity of the sun's rays in the tropics and on high mountains, and surveys of the distribution and variations of air pressure and temperature in North America comparable to what he had done in Europe. Abbot Lawrence Rotch from the Blue Hill Observatory in Massachusetts sought funds for a three-month expedition to fly instrumented kites from the stern of a rented steamer plying the tropical South

Atlantic. Weather Bureau chief Willis Luther Moore supported "general meteorological bibliography," compilation of a daily international weather map, new observations of the upper air and weather over the oceans, and investigations relating weather to terrestrial magnetism, atmospheric electricity, and solar radiation. Frank Bigelow suggested that Carnegie establish a fully instrumented meteorological laboratory in Washington, DC, and a cadre of human computers on call for scientists in need of such "mechanical work." The Department of Terrestrial Magnetism at Carnegie soon got its well-funded laboratory and its worldwide magnetic expeditions, but meteorology did not.[32] Instead, the Carnegie Institution followed Abbe's advice and looked for "the pre-eminent right man."

Bjerknes's Theoretical Program

We shall arrive at the desired result [satisfactory long-range seasonal forecasts] sooner and better by the study of the mechanics of the atmosphere than by the search for elusive empiric periodicities.
—Cleveland Abbe, 1901[33]

In 1895 Cleveland Abbe pointed out that meteorology needed a deductive treatise on the laws governing the atmosphere as complete and rigorous as the *Celestial Mechanics* of Pierre-Simon Marquis de Laplace:

Meteorologists can never be satisfied until they have a deeper insight into the mechanics of the atmosphere. Something more is needed than the most perfect organization for observing, reporting and publishing the latest news from the atmosphere. It is not enough to know what the conditions have been and are, but we must know what they will be, and why so. We must have a deductive treatise on the laws governing the atmosphere as complete and rigorous as the "Celestial Mechanics" of La-Place, and this will necessarily be a treatise on the application to the atmosphere of the general laws of force, or what is technically known as the dynamics and thermodynamics of gases and vapors.[34]

Less than a decade later, Vilhelm Bjerknes initiated a program to measure current atmospheric conditions and to calculate the future state of the weather using the equations of hydrodynamics and thermodynamics. He said this in 1902 in the *Meteorologische Zeitschrift*: "Each task of theoretical mechanics is, when it is placed in direct form, a prognostic, just like the most well-known task of practical meteorology. The goal is to predict the dynamic and physical condition of the atmosphere at a later time, if at an earlier given time, this condition is well known."[35] According to Bjerknes, the central problem of the science of meteorology is weather prediction by rational dynamical-physical methods. He wanted to place meteorology on solid observational and theoretical foundations.

In 1904, Bjerknes published a landmark paper, "Das Problem der Wettervorhersage betrachtet vom Standpunkte der Mechanik und der Physik" (Weather Forecasting as a Problem in Mechanics and Physics), in the *Meteorologische Zeitschrift*. In it, he set out two criteria for preparing an objective forecast of the future state of the weather using the equations of hydrodynamics and thermodynamics:

1. A sufficiently accurate knowledge of the state of the atmosphere at the initial time and

2. A sufficiently accurate knowledge of the laws according to which one state of the atmosphere develops from the other.

The first step he called diagnosis, and the second, prognosis.[36]

If Bjerknes had claimed perfect knowledge of the laws of nature and perfect accuracy in measurement, he would have been following Laplace's famous dictum of universal determinism—that every purely material-mechanical problem can be reduced to stating the present position and motion of all mass particles involved and then predicting their future positions and motion at a given time by the laws of mechanics. In *Mécanique céleste* (1799), Laplace set out to reduce all known phenomena of the world to the law of gravity by strict mathematical principles. In his *Essai philosophique sur les probabilités* (1814), Laplace was more explicit:

We must therefore consider the present state of the universe as the effect of its previous state and the cause of those who will follow. An intelligence which, for a given moment knew all the forces by which nature is animated and the respective situation of the beings who compose it, if moreover it were vast enough to submit these data to analysis, would embrace in the same formula movements of the greatest bodies of the universe and those of the lightest atom: nothing would be uncertain for it, and the future, like the past, would be present to its eyes.[37]

Astronomers experienced great success using this dictum for solving extremely simple cases of periodic, frictionless, two-body motion. Their computational successes elevated their scientific status and their public celebrity, perhaps undeservedly so. But meteorologists, using the same dynamic laws but a vast number and complexity of variables, face a public that demands impossible details of an impossible forecast. There is no superhuman forecasting demon with unlimited observational power and unlimited dynamical knowledge; the perfect Laplacian weather forecast is a chimera.

Bjerknes's use of the term "sufficiently accurate" in his formulation tempered his determinism and undoubtedly stemmed from his exposure to the lectures of Henri Poincaré and to his personal experience in trying to measure, with any precision, the initial state of the atmosphere over an extended area. Poincaré had raised the issue of a computation's sensitivity to initial conditions and undermined the notion of a perfectly precise observation of nature, noting that, "small differences in the initial conditions produce very great ones in the final phenomena." This is now recognized as one of the canons of chaos theory (chapter 5). Nevertheless, Bjerknes pursued a neo-Laplacian, quasi-deterministic program, with real-time weather data and calculations of "sufficient accuracy." This was a major step forward, aligning meteorology with physical law rather than art or craft. It attracted considerable international attention and agreement and garnered Bjerknes considerable fame.

Bjerknes claimed that a light came on when he shifted his focus from the motions of classical ideal fluids that were homogeneous and incompressible to the "impractical" but real fluid systems comprising the atmosphere and the sea. The motions of real fluids depend on differences of density caused by differences of temperature and humidity in the atmosphere and differences of temperature and salinity in the sea. The absence of homogeneity is called *baroclinic*, while the homogenous condition is called *barotropic*, a distinction he learned from Ekholm (see figure 2.1).

We live at the bottom of a baroclinic atmosphere along the shores of baroclinic oceans. The general circulation of the atmosphere and oceans is driven by differences at the surface in the uptake of thermal energy from the sun, poleward tranfers of sensible heat, latent energy released and absorbed by phase changes in the hydrosphere and cryosphere, and deflections due to Earth's rotation. The winds and currents of the globe are never at rest; they circulate endlessly in large and small vortices, called solenoids; the system always seeks equilibrium but never attains it.

On the other hand, in barotropic conditions the density of the fluid is only a function of the pressure, and the forces are symmetric. The atmosphere may approximate this condition when the flow is laminar, unheated, and dry. Seawater is barotropic when its salt content is uniformly stratified;

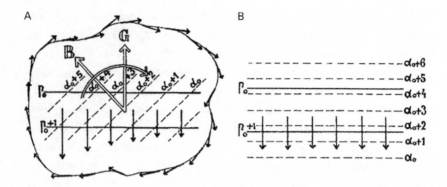

Figure 2.1
Distinction between (a) baroclinic and (b) barotropic fields of mass. The solid lines indicate equal pressure, and the dashed lines equal specific volume or density. Vectors B and G represent, respectively, the density gradient and the pressure gradient. Their cross-product, or curl, indicates the nonconservative rotational force acting within the baroclinic fluid, generating solenoids. Because vectors B and G are parallel in the barotropic fluid, the curl and thus the rotation are zero. *Source:* V. Bjerknes, "On the Dynamics of the Circular Vortex."

winds at high altitudes can approximate barotropic conditions when they have no curvature. In such cases, to a first approximation, there is no mixing and there are no rotational forces to generate circulation or storms.

How were air motions represented on charts and then analyzed? The Eulerian approach is synoptic, focused on air moving past fixed points. An observer with a bird's eye view could apply Euler's equations to the patterns of air movement as recorded at *all* the points. This approach produces wind streamlines depicted on a map. The Lagrangian approach, using equivalent equations in different form, follows a particular parcel or particle of air along its pathway, producing a trajectory analysis. In discussing wind patterns and air motions, we might refer to these depictions as maps and paths. An apocryphal (but true) story involves Leonhard Euler's annual lectures in St. Petersburg, Russia. At the end of each lecture series, Euler ceremoniously invited his students onto the bridge over the Neva River. There the great mathematician would tear his lecture notes into bits and sprinkle them in the water, telling his students to follow them from above and analyze their motions as they swirled in the current. One year, Joseph-Louis Lagrange attended the ceremony as a distinguished visitor and, seeing the moving bits, immediately jumped off the bridge and began swimming along with them downstream. As he turned a bend in the river, Euler and his students heard the Italian prodigy calling to them: "Go with the flow!"

A Visit to America

In 1905, Vilhelm Bjerknes visited the United States to lecture at Columbia University. In addition to New York, he also visited Washington, DC, New Haven, and Montreal. The invitations came from mathematical physicist and electrical engineer Michael Pupin, a student of Helmholtz, Arthur L. Day, director of the Carnegie geophysics laboratory, and his long-time correspondent Cleveland Abbe. In the first three weeks of December 1905, Bjerknes delivered a series of seven lectures at Columbia University on "Fields of Force."[38] He began with Faraday's idea of electric and magnetic fields of force, as developed by James Clerk Maxwell and his own mentor, Heinrich Hertz. He then generalized these results for perfect fluids as he had done with his father. His ultimate goal was to introduce a new, more general method to treat hydrodynamic fields that combined elementary reasoning and experiment, the equations of motion for a perfect fluid in both their Eulerian and Lagrangian forms, and Maxwell's equations for the behavior of electromagnetic fields.

Bjerknes realized the "practical importance" of mastering the dynamics of the two universal media, the atmosphere and the sea, on which humans are completely dependent. He saw no opposition between theory and practice and, in his penultimate lecture at Columbia on December 23, left his audience with the inspiration that the more we advance in theoretical *and* practical research, the more we shall discover. He also invited them to an additional lecture the following day on the hydrodynamic fields of force in the atmosphere and the sea.[39] Here he announced that the problems of meteorology and oceanography—the change of seasons, ocean currents, the onset of the sea breeze, the upwelling of seawater, the origin and dissipation of cyclones—can be understood by the investigation of solenoids.

Bjerknes had sketched out a method based on observations and the laws of dynamics and thermodynamics for forecasting the weather and the future state of the sea. His program was aligned with the quest of classical mechanics for prediction, both to prove theory and to supply public demands. His quest was to organize cooperative observations throughout the world, "an international organization having for its object the solution of the problems of the forecasting of the future states of the atmosphere and the hydrosphere." Abbe was excited, as was indeed the entire audience. The lectures solidified Bjerknes's reputation as the "highest living authority in the study of the mechanics of fluids."[40]

New York City impressed Bjerknes, "more than any other city that I have seen," as did Columbia. He found the pace rather hectic, with numerous receptions and luncheons, courtesy visits to Norwegian acquaintances, and even a football match between Columbia and Penn (the home team lost 23–0). He had a twinge of homesickness as he imagined that Broadway and its crossing streets "seem sometimes almost as the valleys of Western Norway, especially when it gets dark." Bjerknes enjoyed a tea party hosted by atomic physicist Harriet Brooks, Ernest Rutherford's first graduate student and a faculty member at Barnard College. He enjoyed the company and the witty conversation, writing to his wife that this was "one of the funniest and nicest experiences here." He left the party with an invitation from Mrs. Rutherford to visit Montreal. Bjerknes and his friend, assistant professor of physics Albert P. Wills, dined with electrical engineer Peter Cooper Hewitt, "an inventor of high rank, who in the real American way, despises difficulties, both technical and economical," and toured his laboratory at Madison Square. Later that week, they visited the home of Michael Pupin, where after dinner, Bjerknes unfolded his maps on the floor and "developed his great plans." As Bjerknes recalled, "The man is intelligent, he understood me, he lay on the soft carpet and studied the maps while I lectured. So now I have two followers, Wills and Pupin." The following day, Bjerknes visited the New York Weather Bureau office, where he was able to spend time with Cleveland Abbe, "the most charming old man one can imagine, an ageing angel." Bjerknes and Abbe dined together and talked meteorology the whole evening. There Abbe informed him that Weather Bureau chief Willis Moore wanted him to talk to a thousand people in Washington. The balance of his itinerary included visits to Washington, New York (to lecture to the American Physical Society, where he was elected an honorary member), New Haven (to visit mathematician Edwin Bidwell Wilson), Montreal (to see Rutherford), and Boston (to visit Harvard and lecture at MIT).[41]

When Bjerknes arrived in Washington, DC, Carnegie Institution president Robert S. Woodward met him at the train station and took him to Weather Bureau headquarters. Bjerknes confided to his wife that chief

Willis Moore "made a most repulsive impression on me, despite his forceful compliments." He also met Charles Marvin, head of the instrument department, and Frank Bigelow, of whom Bjerknes wrote, "He was a sad sight. He is a complete gentleman, but he is not right in his head." The Weather Bureau was not in the practice of hosting large lectures, so instead Bjerknes lectured at the Carnegie Institution on "Weather Forecasting as a Problem in Mechanics and Physics." He displayed on the wall an "impressive number" of Sandström's charts, including maps that were graphical representations of the fields of force on which the further developments of the motions in the atmosphere and the sea depend (figure 2.2).

The charts Bjerknes displayed in his lecture represented both computational aids and a new way of looking at the weather. Bjerknes connected wind arrows representing the horizontal motion (figure 2.2a) to construct streamlines depicting lines of flow and curves of equal wind intensity (figure 2.2b). From Bjerknes's perspective, streamlines depict the winds at a specified moment in time and provide a snapshot of wind or current velocities over a region. They are constructed by connecting velocity vectors as measured at fixed points. The paths of particles moving through the fluid are called *trajectories*. Thus the air or water will flow, in the next instant, according to the patterns depicted by streamlines, while moving parcels of air or water, subject to changing forces, trace out the trajectories. The synoptic view is Eulerian, focused on air moving past fixed points. Following the trajectory of a parcel of air is Lagrangian. This is the distinction between maps and paths.

Bjerknes pointed out that such charts of motion possess many characteristic features in the form of singularities and lines of convergence, which are in an obvious relation to the conditions of the weather and will lead to practical rules for weather forecasts. Areas of convergence, such as the one depicted in Minnesota, represent areas of strong vertical motion and likely precipitation. Bjerknes's depiction of streamlines on a map, later identified as frontal surfaces, was a visual way to represent and analyze hydrodynamic relations. The charts of motion also had potential application to aerial navigation. For instance, an airship moving from Bismarck, North Dakota, to Lake Superior would not be able to fly in a straight line because it would encounter strong headwinds, so the pilot might decide to travel south of the center of the vortex to minimize travel time. Two streamline charts, separated by an interval of time, would allow for the calculation of accelera-

A

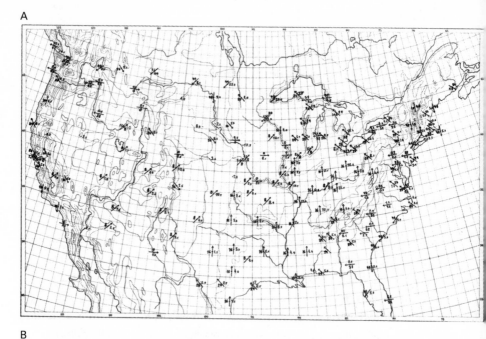

B

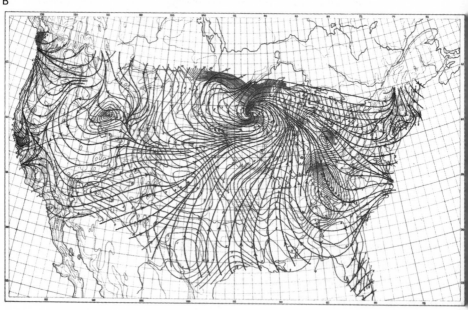

Figure 2.2
(a) Wind observations on a U.S. weather map. Original caption: "Arrows and numbers (heavy) representing wind direction. Numbers (ordinary type) representing wind intensity (m/sec.) U.S.A. 1905, Nov. 28, 8 A.M. 75th Meridian Time." (b) Streamlines. Original caption: "Lines of Flow and Curves of Equal Wind-Intensity (m/sec.). U.S.A. 1905, Nov. 28, 8 A.M. 75th Meridian Time." *Source:* V. Bjerknes and collaborators, *Kinematics*, plates 36 and 38.

tions and would form the basis of a forecast displaying not just the movement of weather systems but also their intensification or dissipation.

The lecture was a big success, "interrupted by applause in the middle when all the hydrodynamic formulae turned out to fit with the electrical ones."[42] As he recalled four decades later, "If ever any lecture of mine made a splash, this was it." The problem, as he depicted it in that lecture, was of colossal dimension. It inspired the Carnegie Institution to provide him with continuing funding and, as Bjerknes reminisced, "this strongly tied me to the meteorological branch of the line of thought that my father had started to develop."[43] Carnegie had found an exceptional man; meteorology had found its new leader.

The First Atomic Age

The "first atomic age" at the dawn of the twentieth century rocked the foundational assumptions of classical physics and introduced the world to roentgen rays and Becquerel rays (later called X-rays and radioactivity). Bjerknes likely discussed these issues with Rutherford during his visit to Montreal in 1905. These emanations were first seen as mysteries, then as commercial products, and soon as "the icons of a new and modern science" in all its chaotic, fantastic, and ultimately threatening dimensions.[44] In the first two decades of the twentieth century, physicists speculated that radioactivity might be the cause of the ionization of the atmosphere and might play active roles in meteorological processes by providing an energy source large enough to influence the stability of a column of air or powerful enough to maintain hitherto unknown electrical circuits in the atmosphere.

In 1901, a pair of German physicists, Julius Elster and Hans Friedrich Geitel, measured the ionization of the air and its spatial and temporal variations, which they assumed was due, in part, to the decay of radioactive elements in Earth's crust.[45] Several years later, working independently, physicists H. A. Bumstead and Arthur S. Eve suggested that radioactivity detected in rain and snow was probably due to radium-excited activity and was sensitive to the height at which the raindrops formed.[46] In his 1905 lecture to the Nobel Prize Foundation, Pierre Curie opined that radioactivity probably plays an important role in meteorology, hypothesizing that the ionization of the air provokes the condensation of water vapor.[47]

The community of atomic physicists gave these matters considerable attention, yet there was little crossover to scientists in other fields, including meteorology. Svante Arrhenius's book *Worlds in the Making* contains only a few brief, speculative, yet insightful comments on radium. He pointed out that a gram of radium produces a quarter of a million times more total heat than the combustion of a gram of carbon and may be a possible source of the sun's energy. He also speculated that radium might absorb radiation coming from space, in some unknown manner.[48] Following a long series of careful observations taken by Victor Franz Hess on and under the sea, on the Eiffel Tower, and with instrumented balloons, the scientific community concluded that ionizing radiation was penetrating the atmosphere from outer space.[49] Robert Millikan named the phenomena "cosmic rays" in 1926, and Hess won the Nobel Prize for this work in 1936.[50] There is no evidence that Bjerknes studied questions of the meteorological implications of atmospheric ionization raised during the first atomic age.

The Carnegie Proposal

Shortly after returning home from his American trip, Bjerknes filed his first application for research support from the Carnegie Institution, "To prepare a scientific work on the application of the methods of hydrodynamics and thermodynamics to practical meteorology and hydrography."[51] Galileo had conducted dialogues on two new sciences—strength of materials and kinematics. Bjerknes also was aiming to establish two new sciences—dynamic meteorology and dynamic oceanography. The time to attempt this had now arrived.

In his research proposal, Bjerknes explained that prediction of the weather is the "great problem of meteorology." Stated in mathematical form, it involves the solution of seven independent equations to calculate seven quantities—velocity (in three dimensions), pressure, density, temperature, and humidity. Using either Eulerian or Lagrangian analysis, the equations are based on hydrodynamical motion in three dimensions, the conservation of mass, the properties of an ideal gas (equation of state), and the principles of thermodynamics. Aligning his program more with Laplacian certainty than with Poincaré's indeterminacy, Bjerknes naively thought the problem was "perfectly determinate already in our present state of knowledge." He appeared to have forgotten, or at least omitted, his notion of a *sufficiently* accurate forecast.

For Bjerknes, the path forward involved measuring the initial state of the atmosphere "with tolerable completeness" using a dense network of surface stations and upper-air observations collected by kites and balloons. The solutions of the seven equations would describe the future state of the atmosphere. An analogous problem was ocean forecasting, a problem of great practical importance because fisheries respond to the changing state of the sea much in the same way that agriculture responds to the changing states of the atmosphere. Because the atmosphere and sea are linked by energy and moisture exchanges, Bjerknes thought it rational to treat both forecasting problems side by side, emphasizing that "it is the right time to attempt this." Bjerknes had been at work for several years on a book that introduced dynamical methods into practical meteorology and hydrography and that would provide graphical or mechanical forecasting methods "in a fully worked out form, ready for practical use."

Bjerknes included his collaborator, Johan W. Sandström, a practical meteorologist and hydrographer, in his Carnegie Institution proposal. Sandström was born in northern Sweden in 1874 and first worked in Stockholm as a factory workman. Later he attended lectures on mathematics and mechanics at Stockholms Högskola and joined the meteorological office and then the hydrographic service. Bjerknes proposed using Carnegie funding to free Sandström from his official work and allow "a man of extraordinary scientific value the opportunity to distinguish himself, and thus possibly, through the value of his work, to acquire a position enabling him to devote himself entirely to scientific work."[52] Bjerknes received a first grant of $1,200 in 1906 (about $30,000 in 2016 when adjusted for inflation) and hired Sandström and an assistant, Bengt Söderberg. Bjerknes used his grant money to support his young assistants and advance their careers. He moved from Stockholm to Kristiania in 1907 as a professor of applied mechanics and mathematical physics and worked with his assistants Sandström, Olaf Devik, and Theodor Hesselberg to produce the first two volumes of *Dynamic Meteorology and Hydrography*—volume 1, *Statics* (1910), and volume 2, *Kinematics* (1911).[53]

In 1910, at the invitation of British meteorologist Sir Napier Shaw, Bjerknes brought his plea for "observations which could be or which deserved to be treated according to rational dynamical methods" to members of the Royal Meteorological Society assembled at University College London.[54] The biggest challenge that Bjerknes faced in advancing his theoretical program was getting high-quality weather data that could be subjected to vector analysis. In his mind, it was like an astronomer's need for accurate observations of heavenly bodies to calculate orbits.

In Transit, Leipzig 1913–1917

There is after all but one problem worth attacking, viz., the precalculation of future conditions.
—Bjerknes, 1913[55]

In 1912, Bjerknes accepted a call to the University of Leipzig, where he would direct the newly organized Geophysical Institute, work with practicing aerologists who were adopting his methods, and combine his research and teaching. He felt constrained in Norway, lecturing in a language understood "by only a very limited number of students of science."[56] Hesselberg and Harald U. Sverdrup followed him to Leipzig to work on frictional forces, but his family stayed behind in Norway. There they tried to complete volume 3, *Dynamics*, but were frustrated by the complexity of the undertaking, the lack of comprehensive data and a useful model, and the confusion caused by the world war. They did, however, produce novel three-dimensional synoptic representations of atmospheric states using data collected during cooperative campaigns to measure the upper air—so-called international aerological days—held between 1900 and 1913.

On January 8, 1913, Bjerknes presented his inaugural lecture, "Die Meteorologie als Exacte Wissenshaft," in the aula of the university. He began with a nod to geophysics, conceived as a part of cosmical physics and divided into three branches—physics of the atmosphere, physics of the hydrosphere, and physics of the rigid earth. His research program included the first two. The lecture reviewed the history of meteorology from Aristotle to recent developments in aerology and recent progress in formulating the seven equations for the seven variable meteorological elements. Bjerknes announced that the ultimate goal of his program was to "apply the equations of theoretical physics not to ideal cases only, but to the actual existing atmospheric conditions as they are revealed by modern observations."[57] Facing the daunting and seemingly intractable analytical problems this would entail, Bjerknes previewed the techniques that ultimately led to the Bergen school methods. He proposed presenting observations on charts and recasting computations as graphical mathematics, which would allow one map to be derived from the other, "just as one usually derives one equation from another." He was undaunted by the time it would take to prepare a forecast because he was most interested in proof of concept, with practical results to be developed later. His ending was memorable: "It may require many years to bore a tunnel through a mountain. Many a laborer may not live to see the cut finished. Nevertheless this will not prevent later comers from riding through the tunnel at express-train speed."[58]

In May, Bjerknes sent Carnegie president Woodward the first issue of a new publication, the *Veröffentlichungen* of the Geophysical Institute of the University of Leipzig, aimed at applying the methods contained in the first two volumes of *Dynamical Meteorology and Hydrography*. He considered the publication "a direct fruit" of Carnegie support and promised to send additional copies as they appeared. He reported success in getting the International Meteorological Committee at its recent meeting at Rome to recommend that all aerological publications employ the centimeter, gram, second (CGS) unit of pressure—the millibar. He hoped that the British, Americans, and Europeans would soon follow and abandon inches and millimeters of mercury as pressure units.[59] Woodward encouraged him with news that a change in the administration of the Weather Bureau was coming soon, "and we are hoping very much that we may secure a bureau chief who may have something like an intelligent and scientific knowledge of meteorology as well as of administration."[60]

At the U.S. Weather Bureau, Willis Moore had been fired as chief and Charles F. Marvin was in. On January 1, 1914, Marvin's bureau began the welcome practice of issuing daily weather maps of the Northern Hemisphere using CGS units and nomenclature, with data collected by modern systems of telegraphy, cable, and wireless.[61] Marvin, who had risen through the ranks to become a leader in the development of instruments, specifically research kites, hoped that the new map would facilitate and promote the serious scientific study of the great and complex problems of aerology and the general circulation of the atmosphere. He wrote that he was "following the suggestion of Bjerknes" in using the absolute units of pressure and temperature. Bjerknes wrote back from Leipzig, "I have been pleased to see the publication of your daily weather map for the Northern

Hemisphere. The introduction of the C.G.S. units is a very great progress indeed."[62] The new units simplified the equations of motion and supported the move, already in progress, from statistical climatology to dynamic meteorology in which observations of pressure were taken not for their own sake but to compute from them forces and accelerations, the energy of the system, and the heat evolved. The British Meteorological Office also offered praise, with Sir Napier Shaw writing, "One of the striking features of the maps now issued by the Weather Bureau is that for the first time in the history of official meteorological institutions C.G.S. units of pressure and the absolute scale of temperature are used for a daily issue of charts. . . . This is indeed a remarkable step toward the unification of the methods of expressing pressure over the globe."[63] Cleveland Abbe thought it was a great step forward for the Réseau Mondial and for Bjerknes's program of dynamic meteorology. The daily maps would allow for diagnosis and prognosis of the atmospheric streamlines—the lines of force and the lines of flow. Eventually, Abbe hoped, the seven fundamental factors that control atmospheric motion, gathered into one set of systematic equations or graphic charts, would fully explain the phenomena depicted on the maps.[64] Standing behind this new initiative, with his name and the U.S. Department of Agriculture Weather Bureau proudly emblazoned across the top banner (figure 2.3), was Weather Bureau chief Charles Marvin. The hopes of meteorologists were dashed less than seven months later when the opening of hostilities in Europe resulted in the complete suspension of the telegraphic transmission of foreign reports and ended the issuing of the globe-circling maps.[65]

On July 29, 1914, one day after Austria-Hungary declared war on Serbia and three days before Germany entered the war, Bjerknes sent an optimistic report to the Carnegie noting that the work in Leipzig was proceeding "very well." Bjerknes had published volume 3 of his *Veröffentlichungen* series containing synoptic representations of atmospheric conditions—sea-level pressure, temperature, streamlines, isotachs, cloud cover, precipitation, and upper-air measurements at ten levels—over Europe for May 18–20, 1910, from observations taken during international aerological days. Bjerknes was applying his diagnostic and prognostic methods to these charts. His assistants Hesselberg and Sverdrup were with him, although they were not officially connected with the institute. Several dozen introductory students were learning the techniques of kinetics and kinematics, and seven

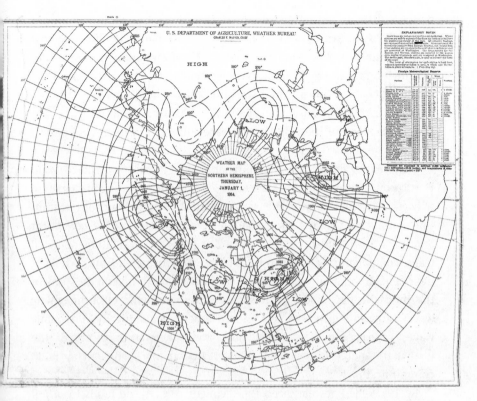

Figure 2.3
Polar projection weather map of the Northern Hemisphere, January 1, 1914, showing multiple weather systems. *Source:* National Oceanic and Atmospheric Administration.

graduate students were working on doctoral degrees.[66] A fortnight later,
Bjerknes's tone had changed significantly: "I hope that the most unnecessary and most cruel of all wars shall not disturb the work which I am performing for the Carnegie Institution." He was planning to go to Bergen in
late August to meet with Helland-Hansen and his oceanography group and
then return to Leipzig in mid-October. He worried about how he would pay
his assistants and whether money could be transferred from Germany to
Norway during the war. He suggested that his assistants might perform
their work in Washington, DC, "On account of the enormously rich collections of meteorological observations stored at the U.S. Weather Bureau,
and the high quality of these observations, many investigations can be

performed there only and nowhere else in the world." He thought the Weather Bureau would also profit from this exchange but worried that transatlantic passage was no longer safe.[67] Woodward approved of the idea in principle and also made immediate arrangements to pay the assistants in Bergen in case the Leipzig funds were unavailable.[68] Another pressing concern was that war would impede scientific progress and diminish the purchasing capacity of the world's monetary currencies, which would affect the Carnegie grants.

Bjerknes's earlier work in electrodynamics and his strong connections to Hertz provided him with intimate knowledge of the operating principles of wireless telegraphy. During the war, he visited the large Marconi transmitting station built by the Norwegian Telegraph Service on the coast near Stavanger. He found it "amusing" to discover that the station operated using components derived from a Hertzian oscillator and his own resonance circuit, built on a giant scale.[69] He might have mused on the irony of the move he had made, professionally, from studying electromagnetic waves propagating in the old idealized ether to studying hydrodynamic waves propagating in the real fluid atmosphere and oceans of the planet, the global study of which would be greatly facilitated by radio communications.

By 1915, Bjerknes had been the recipient of Carnegie funds for a full decade. He used the occasion to reapply for funding and to summarize his accomplishments to date and his near-term vision of the future. The war had not diminished the team's scientific activity: Hesselberg and Sverdrup, being Norwegians, had "nothing to do with the war." Of the few students studying at the institute, one was his second son, Jacob Bjerknes, who "is very much interested in this branch of science . . . though he is still too young to devote himself to this work [and] will have to occupy himself principally with his general studies." In his request for continued funding, Bjerknes wrote about his past accomplishments in publishing the two diagnostic volumes, *Statics* and *Kinematics*, the general but not yet official acceptance of the CGS system of units, and the grant that allowed four young men to get their start in scientific research. Hesselberg and Sverdrup were still working with him, Sandström was employed by the Swedish Weather Bureau in Stockholm, and Devik was tasked with creating a regular weather service in northern Norway. The primary goal remaining was to complete the final volume, which would focus on prognostics, but the going was slow. Friction and stability criteria were tough nuts to crack, but Bjerknes promised he would devote himself "entirely" to the redaction of volume 3, "I hope without interruption until I have finished it."[70]

Bjerknes persevered through thick and thin, mostly thin, hoping the war would come to a quick resolution. It did not. Bjerknes and his assistants summered in Norway and returned to Germany each autumn, leaving their families behind. Leipzig was situated far from the theaters of war, and superficially, conditions were "almost normal." Yet things were not at all normal. Army conscription had greatly reduced the numbers of matriculated students in Leipzig, and Bjerknes sent his international correspondence unsealed and written in German for the convenience of the government censors.[71] He had little direct involvement in the conflict, unlike his Austrian colleague and bête noire, Felix Exner, who commanded the German weather services. Still, the German military made widespread use of Bjerknes's work, and the field weather services consulted him regularly in support of aviation, artillery shelling, and poison gas attacks. Herbert Petzold, a German doctoral student who had been studying convergence lines in the wind field, was killed at Verdun in 1916. Jacob Bjerknes took over his research, assisted by Halvor Solberg, and published his first scientific paper on the movement of convergence lines, reporting that they may be thousands of kilometers long, tend to drift eastward, and are connected with clouds and precipitation.[72] The conflict ended the promising international mapping project and drove Bjerknes and his team to Norway after the deprivation of the *Kohlrübenwinter* (turnip winter) of 1916–17. In the absence of the world war, what we know today as the Bergen school of meteorology might never have developed, or it might have been called the Leipzig school, based on the analysis of streamlines and built on a dense network of German rather than Norwegian stations.

Bergen and the Turn to Forecasting

I am now back again in Norway, not merely for a summer journey, but forever.
—Vilhelm Bjerknes, 1917[73]

In 1917, Vilhelm Bjerknes left war-torn Germany to accept a professorship at the new *Geofysisk institutt* established by his friend Bjørn Helland-Hansen at the venerable Bergen Museum,. Bergen had a long tradition as a center of scientific activity, and the long Norwegian coastline was an excellent place for all kinds of geophysical work. Bjerknes was delighted to be back again in his own country and sought to focus his energies on his meteorological task of "enormous dimensions." The first order of business was to convene a strategic planning conference "in order to draw up the great lines for these investigations" and sketch the outlines of a Norwegian geophysics association. Bjerknes scheduled the meeting in August 1917 at his summer residence at Geilo, a delightful mountain village located halfway between Oslo and Bergen. Bjerknes was joined by Helland-Hansen and former collaborators Sandström, Hesselberg, Devik, and Sverdrup, among others. The meeting, emphasizing linkages between the theoretical and the practical, resulted in a new fund for geophysical research named in honor of Kristian Birkeland, a new multilingual journal, *Geofysiske publikasjoner*, and outreach to Norway's industrial, commercial, and military leaders.[74]

Bjerknes made a radical career turn into practical forecasting to "assist the production of breadstuff," which had become an urgent national problem. He took the initiative, early in 1918, of "introducing an extended weather service," the *Vervarslinga på Vestlandet*, that would combine theoretical and practical forecasting methods during the three critical summer months, July to September, with Tor Bergeron, Solberg, and his son Jacob serving as forecasters.

On August 28, 1918, at the Congress of Scandinavian Geophysicists in Göteburg, Sweden, Vilhelm Bjerknes presented a keynote address on "Weather Forecasting." He outlined two special challenges he had faced for two decades—the practical (obtaining the necessary data through observation) and the theoretical (developing the methods and applying the equations to the observations). Still, without a method of solving the complete mathematical problem in the near future and thereby untying the Gordian knot of forecasting, Bjerknes focused his efforts on historical methods of map construction and graphical analysis. He reminded his audience that mapping streamlines were important for indicating the lines of flow and the strength of the wind, that he and his collaborators had deduced from them lines of convergence and divergence, and that his son Jacob had connected this all to the formation and dissipation of cyclones.[75]

Jacob Bjerknes's classic paper of 1919, "On the Structure of Moving Cyclones," concluded by paying homage to a long line of predecessors, including Heinrich Wilhelm Dove's theory of the conflict between polar and equatorial currents; William Ferrel's convectional theory, confirmed in essence by air rising into the cyclone from the warm sector; and Julius Hann's view that cyclones were local manifestations of the general circulation of the atmosphere. The Bergen school established a network of surface stations across western Norway that was some ten times denser than what they had used in Leipzig (figure 2.4a). This was supplemented by methods of indirect aerology based on the motion and character of clouds, visibility, and the amounts and types of precipitation. The Bergen school revolutionized the synoptic map by adding topographic details, including observations from every station, applying tools of objective analysis to the observations (figure 2.4b) and depicting streamlines as an instantaneous approximation to trajectories. Analysis of the streamlines helped in the identification of two kinds of lines of convergence—the squall line and the steering line—later labeled warm fronts and cold fronts (figure 2.4c).[76] In his review of the history of scientific weather analysis and forecasting, Bergen meteorologist Tor Bergeron applied what he called "O, T, M" (observation, tools, method) analysis to the subject, using O to represent meteorological observations, T the tools (whether conceptual, graphical, mathematical, or mechanical) for evaluating the data, and M the models of atmospheric structure and their degree of rational physical approach to the problem. All three elements were necessary for complete theory and practice.[77] For the Bergen school, this consisted of a network of surface and aerological observation (O); a new synoptic weather map featuring streamlines, areas of precipitation, and realistic topography serving as a diagnostic tool (T); and a new prognostic model (M) of atmospheric structure with fronts, air masses, and developing cyclones.

The main accomplishment of the Bergen school lay in its use of realistic models of structure that provided rules for steering and intensification of storms and thus were directly connected to the weather forecast. They conducted classical case studies of a cold front (July 24, 1918) and an occluded front (November 18, 1919), in which the cold- and warm-air sectors are separated. From such studies, they developed the life-cycle model of cyclone development. Overall, their aim was to minimize the subjectivity of forecasting by using explicit and physically reasonable rules of map analysis and evaluation of upper-air conditions. Such methods were aimed at turning every new weather analysis, whether for research or forecasting, into a creative moment of discovery—not to replace intuition but to enhance it.[78]

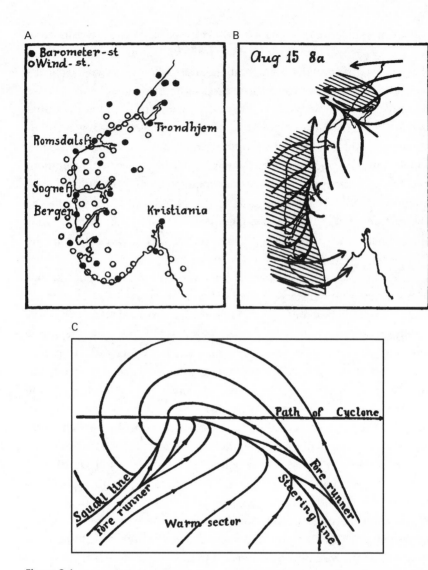

Figure 2.4
Tor Bergeron's "O, T, M" (observation, tools, model) in 1918. (a) Observation stations arrayed in a dense network over Norway; (b) tools of analysis depicting streamlines and shaded areas of precipitation; (c) model of an open-wave cyclone.

The German military had been interested in Bjerknes's forecasting methods, but the enthusiasm was not mutual. His turn to practical forecasting was motivated by the food crisis in Norway. Fishermen expressed their gratitude for the help that the storm warnings provided in their hard and dangerous work. To emphasize the economic value of the forecasts, Bjerknes recounted a recent successful forecast for a farming community issued in the morning for rains that afternoon: "Alarmed by the forecast, the peasants began, in spite of the bright weather, to fetch in the hay even in the forenoon, and just in time, as they afterwards saw." He celebrated the fact that the help had come visibly and directly from science to the working people, "and not by the . . . long way about which make common people forget that *science* was the origin of the progress."[79]

Bjerknes summed up the results of the new Norwegian Weather Service in the *Carnegie Annual Report* for 1919:

The atmosphere is crossed and re-crossed by surfaces of discontinuity, separating from each other masses of air having more or less different velocity and different physical properties, showing themselves by differences of temperature and humidity. . . . Almost every change of weather is due to the passage of a surface of this kind. The rain falls from the warm air when it is forced to mount the slanting surface separating it from the underlying heavier and colder air. In the cyclones this takes place on a gigantic scale along the "steering surface" and the "squall-surface." . . . These results give a physically simple and intelligible view of the phenomena of the weather chart and seem promising from a practical point of view. For promoting weather forecasting it will be of high importance to arrange the observations so that the formation of the discontinuities can be detected at an early state, and their propagation followed as accurately as possible on the weather charts.[80]

Bjerknes imagined a similar service for the "improvement of weather forecasting in [such] a great corn producing country as the United States." Launching a trial balloon of his own, he informed Woodward that he was willing to come to Washington for a short or long time with at least one of his assistants to institute such reforms in the U.S. Weather Bureau.[81] Woodward was very supportive, offering to consult with Weather Bureau chief Marvin and seek support from the Carnegie Trustees. He asked Bjerknes for an outline of the proposed research and an indication of the amount of funding needed.[82] When Woodward received the requested information, he turned it over to Marvin at the Weather Bureau, who welcomed the material and had it published in the *Monthly Weather Review*.[83] Bjerknes argued that the most important step for improving weather forecasting in the

United States would be to construct good charts representing the lines of flow, or streamlines, from which cyclone steering lines and squall lines could be identified. He cross-referenced two relevant papers in that same issue—his recent Göteburg address and his son's paper on the structure of moving cyclones—and encouraged forecasters to gain experience using these charts. What made this proposal a nonstarter for the Weather Bureau was Bjerknes's advocacy for a ten- to fifteenfold increase in the density of the network of telegraphic stations in the United States—from the current number of 300 to 4,500. Even if each station required only basic equipment such as a wind vane and thermometer, staffing them and connecting them to the telegraph network made the proposal prohibitively expensive. His suggested compromise—an experimental network along the West Coast or Gulf Coast designed to intercept incoming cyclones—was never implemented.[84]

To promote the accomplishments of the Bergen school and possibly revive the series of daily maps of the Northern Hemisphere that had been interrupted by the war, Bjerknes proposed a polar-front weather service extending around the earth. He argued that the meteorological events of the temperate zone are, in the most intimate way, related to the polar front and its motion. A regular survey was "merely a question of organization" and would aid in the long-range forecasting of cold air outbreaks and in tracking, even over the oceans, families of cyclones and the anticyclones that form when they die (figure 2.5).[85]

In 1921, in a letter to Woodward, who had just retired, Bjerknes looked back on the twenty-three years that had passed since the publication of his circulation theorem and his fourteen years of Carnegie support. The formal task he had undertaken—to work out theoretical methods for the investigation of atmospheric and oceanic phenomena based on the principles of hydrodynamics and thermodynamics—had been "totally changed" in the course of the last three years. When Bjerknes and his group turned to practical weather forecasting, they gained new empirical insights into the dynamics of cyclones, anticyclones, and the general atmospheric circulation, and this new knowledge was now inspiring and opening up new mathematical investigations of the subject.[86]

By taking into consideration the rotation of Earth and by adding a frictional component to his equation, Bjerknes was able to link the Bergen school cyclone theory to planetary-scale horizontal and vertical flow

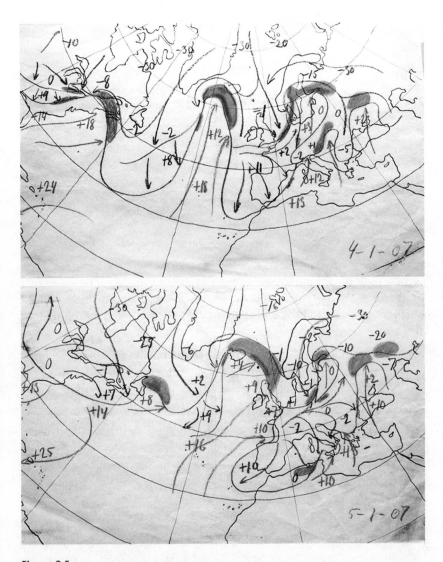

Figure 2.5
Manuscript maps of cyclone families propagating along the polar front on two successive days, January 4, 1907 (top), and January 5, 1907 (bottom). Originals are in color with cold air in blue, warm air in red, and areas of precipitation in green. *Source:* Statsarkivet i Bergen.

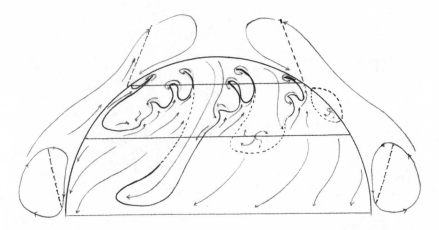

Figure 2.6

A Bergen school view of the general circulation of the Northern Hemisphere as drawn by Anne Louise Beck. Original caption: "Air Circulation of the North Temperate (sic) Zone according to the Bjerknes Theory," from Beck, "Application of the Principles," 21; compare to V. Bjerknes, "On the Dynamics of the Circular Vortex," fig. 31.

patterns. He had learned that contracting circulation loops develop cyclonic rotation and expanding loops develop anticyclonic rotation. He also ventured a theoretical explanation of the Hadley circulation, the trade winds, and the westerlies, in that a vertical loop that becomes inclined acquires an anticyclonic circulation when projected onto a horizontal plane. In mid-latitudes, the increase of the westerly component of the wind with height explains why the lower layers must move poleward and the upper layers equatorward, as they do in the inverse or so-called Ferrel circulation.[87] Depicted in figure 2.6, as components of the general circulation, are three families of cyclones driven by deep cold-air outbreaks along the polar front; a subtropical high-pressure zone; depictions of the trade winds and the westerlies; the Hadley, Ferrel, and polar cells; and sloping vertical surfaces of discontinuity.

Calibration Points and Flash Points

In the summer of 1920, to promote the polar-front theory and the idea of a circumpolar weather service, Bjerknes and his associates at the Geophysical Institute hosted separate conferences, in English and in German, that served as calibration points and flash points for the reception of his theories. Bjerknes hoped to reunite the meteorological community after the war, but the newly formed International Research Council explicitly excluded the participation of scientists from the former Central Powers—thus the need for two conferences. The British delegation, led by Sir Napier Shaw, director of the Meteorological Office, arrived on Monday, July 19. He was accompanied by Ernest Gold, C. K. M. Douglas, and the British mathematician Lewis Fry Richardson, among others. The Bergen Steamship Company and the Bergen-America Line, which took regular observations for the *Vervarslinga på Vestlandet*, provided their transportation from and to Newcastle. For most of the week, they met with representatives from France, Iceland, Norway, Sweden, and the United States for lectures, discussions, map analysis, and the exchange of ideas.

In his opening address to the conference, Vilhelm Bjerknes acknowledged his debt to the ideas of Dove, Helmholtz, Max Margules, and Shaw, among others, and reported on the forecasting success of his group the previous winter using the newly coined concept of the polar front. He outlined the theory of hydrodynamic waves propagating along a surface of discontinuity and emphasized the need for better observational networks. Jacob Bjerknes followed his father, discussing the role of the steering line, the squall line, the warm front (formerly called the anaphalanx) and the cold front (formerly the kataphalanx), the discovery of families of cyclones, and the notion, in need of observational confirmation, that the polar front was a global phenomenon extending to the top of the troposphere. On Sunday and Monday, July 25 and 26, Helland-Hansen led a sightseeing excursion by rail and motorcar to Voringfossen, a 530-foot waterfall at the head of the Hardangerfjord. The participants cruised on the oceanographic vessel *Armauer-Hansen* and were introduced to Norwegian methods of research in physical oceanography.[88] Guests from across Europe arrived on July 25 for the International Commission for the Scientific Investigation of the Upper Air, with Vilhelm Bjerknes in the chair.[89] This was followed, in early August, by a second meeting with German and Austrian meteorologists. Here the sparks flew.

Felix Exner and Vilhelm Bjerknes held conflicting views on dynamic meteorology that they never resolved. Exner devoted two full paragraphs of his brief foreword in *Dynamische Meteorologie* (1917) to attacking Bjerknes. The book presented Exner's survey of the past thirty years of developments in the field and a report on ongoing work. In it, he wrote that Bjerknes had not given sufficient credit to his contemporaries and predecessors, especially Margules. This accusation, repeated by Hermann von Ficker and others, echoed through the decades. As late as 1980, Werner Schwerdtfeger wrote of Bjerknes's "spectacular disregard" of the work of the Vienna school meteorologists and his "unnecessary and unprovoked lack of courtesy" toward his Austrian colleagues.[90] Exner also wrote that Bjerknes and his school had overemphasized the forces of movement in the horizontal plane to the neglect of thermal factors, again a thread that traces back to Margules.

Bjerknes invited Exner to the 1920 conference in Bergen in an attempt to find reconciliation and perhaps heal old wounds, yet he could not resist taking a victory lap based on the recent popularity of the new polar-front theory. Exner's lecture, written in advance, traced his own cyclone theory to the work of Helmholtz, Bigelow, Margules, and Ficker, perhaps to remind Bjerknes of his own lack of acknowledgments. At least the two men broke bread together in Bergen. Exner's published version of this lecture included a few accommodating remarks about parallels with the Bjerknes's theory.[91] Yet in the second edition of *Dynamische Meteorologie* (1925), Exner renewed the attack by printing the foreword to the first edition and providing details of the "irreconcilable differences" between his theory and the Bergen school.

Several scholars have offered interpretations of the conflict of what seemed at times like a war between the Austrian and Scandinavian schools. Deborah R. Coen emphasizes intellectual differences and cultural animosities. She contrasts Exner's enlightened and liberal approach and his tolerance for uncertainty with the dogmatism of his naive "competitors"— cosmopolitan Vienna versus provincial Bergen. Ficker accused Bjerknes of using the "weight of his personality" and a muscular rhetorical campaign employing military metaphors to foist a memorable, if rather sketchy dynamic model on the world—"a schematic hypothesis" that ignored the true nature of atmospheric phenomena.[92] The Vienna school referred to the behavior of air masses using images of freedom, contingency, and independence—not constraint, determinism, and battle, as did the Norwegians. The Austrians considered their work to be original and credible—if less widely accepted then at least "closer to the truth" than the Bergen school theory.

After World War I, the Vienna school's preeminence was severely challenged by developments in Norway. This did not sit well with Exner and his colleagues. Ironically, as Austrian meteorology lost almost all its former advantages of scale after the collapse of the Hapsburg empire, "intriguingly, it flourished because Austrian meteorologists learned to shift strategically between global and local perspectives." They explained this by focusing on the weather in the complex mountainous terrain of Austria, which supposedly "affords more profound insights into weather phenomena," even though Exner developed his theory on his world travels during the greatest reach of the Austro-Hungarian empire.[93]

Robert Marc Friedman viewed the conflict as a manifestation of status anxiety. He attributed Exner's accusations not to any specific scientific disagreements or priority disputes but to stress caused by the "total defeat and dissolution of the Empire, the economic catastrophe that all but shut down scientific work, the international boycott, and the recognition that for the moment Vienna had ceased being a leading center for meteorology." Exner apparently behaved himself in Norway, but on other occasions his "ugly, enraged, and intemperate" verbal outbursts against the Bergen methods were "especially troublesome" and offensive to the "hot-tempered" meteorologist Erik Palmén from Finland and to some degree Bergeron. [94]

Regarding theoretical differences, Exner criticized Bjerknes for restricting his analysis to movements and forces in the horizontal plane, to the neglect of thermodynamics, and also for rejecting the claim that the causes of weather changes at the earth's surface had their origin in the lower stratosphere. The Vienna-Frankfurt school held that warm- and cold-air mass advection in the stratosphere explained the rise and fall of pressure at the ground and thus everything else we call weather systems. The Bergen school countered that, rather than neglecting thermodynamics, their analysis began with the dynamics of differentially heated currents in the troposphere and gradually, with better upper-air measurements, it evolved to include the vertical dynamics of the troposphere and lower stratosphere. It was an argument about the "seat" of atmospheric activity, with the Scandinavians moving upward and the Germans, who had initially been dazzled by the discovery of the stratosphere, starting to move downward.

A mitigating factor may have been the popularity of Bergen school ideas in the Soviet Union and their subsequent reception in Germany. From 1930 to 1932, Bergeron taught Bergen methods in Moscow. On the first day on the job, he met his wife, Vera Romanovskaya, a map analyst in the weather service, who also translated his lecture notes from German to Russian. Bergeron's seminars were well received. The University of Oslo awarded him a doctorate in 1928, but publication of the thesis was delayed until 1934, when it appeared in Russian. Here he discussed the source regions and evolution of different air masses—equatorial, tropical, polar, and arctic—and developed an indirect aerology based on observations of precipitation, clouds, and visibility. It was, in effect, the foundation of a dynamic climatology.

In 1934, one of Bergeron's students, Sergei Petrovich Khromov, published, in Russian, the first complete textbook on Bergen methods. Critics agreed that it filled an important gap in the meteorological world literature, but it was not widely accessible to many readers. Hermann von Ficker supported the idea of a German edition, which was completed in 1940.[95] This edition, *Einführung in die synoptische Wetteranalyse*, edited and translated by Gustav Swoboda, was heavily influenced by Bergeron. He was involved in the German translation and provided important comments on the text and even some original drawings of his own. Topics include the synoptic method, air movements, water in the atmosphere, air masses, fronts, frontal disturbances, and practical forecasting, with a large number of case studies illustrated by maps. Ficker kept a close eye on the publication process to ensure that his work and the work of the Vienna school were properly credited. Now German-speaking meteorologists had a textbook on Bergen school methods. The two warring factions met halfway, symbolically around the 500-millibar level, precisely where Rossby's analysis flourished. According to meteorologist Anders Persson, the meteorological war ended when Hermann Goering told the German meteorologists to accept the Bergen school ideas, just as a new war started.[96] Further reconciliation came after World War II, facilitated by the work of Carl-Gustaf Rossby in Sweden (chapter 3).

Probing the Ionosphere

According to historian Sungook Hong, technology "opens new possibilities and closes some old ones. Whenever this happens, people are forced to think, choose, exploit, and adapt to these new possibilities."[97] Because meteorologists already had seen their field transformed by the electric telegraph in the nineteenth century, they were quick indeed in thinking, choosing, exploiting, and adapting to the new possibilities of wireless and the newer radio technology. By 1914, a technological shift from spark to continuous wave transmission was underway, if not yet complete, and the military communication needs of the combatants came to the fore. This was the year U.S. naval stations began broadcasting abbreviated weather bulletins.[98] The transition from radiotelegraphy (coded messages) to radio-telephony (voice messages) resulted in a dramatic increase in both band-width and public accessibility. A "radio boom" swept the United States in the early 1920s.[99] In 1921, only a handful of stations issued Weather Bureau reports and forecasts, advisories, and warnings in Morse code; one year later, ninety-eight stations in thirty-five states were broadcasting this infor-mation verbally using radiotelephony.[100] That year, the U.S. Weather Bureau began exchanging observations from forty U.S. stations with France "to be broadcast by radio from the Eiffel Tower for the benefit of the European meteorological services within its range." Heady stuff indeed.[101]

In the nineteenth century, scientists who were seeking to explain mag-netic fluctuations at Earth's surface and auroral displays over the poles sug-gested atmospheric electrical currents as a cause. Experiments in radio transmission led Oliver Heaviside to postulate the existence of a reflecting layer in the upper atmosphere that bounced the signals over the horizon. Bjerknes's network in atmospheric physics included Heaviside, whom he befriended during hard times, and extended vicariously to G.F.C. Searle and Edward Victor Appleton.[102] In this way, Bjerknes maintained personal con-nections with electrodynamicists with interests in Earth processes. As infor-mation of an indirect character about the refraction and scatting of radio waves started to proliferate, scientists began discussing multiple layers of ionized gases in the upper atmosphere. In 1926, Robert Watson-Watt pro-posed the name *ionosphere* for these layers, which he defined as a region containing sufficient populations of free electrons with mean free paths sufficient to influence the propagation of electromagnetic waves in com-plex ways.[103] After the multiple layers of the ionosphere were identified, scientists began to explore the ways in which the ionization in the reflect-ing layers varied through the day, through the season, and over the years. The possibility existed of forecasting "ionospheric weather," or the electri-cal state of the upper atmosphere.[104] Some dynamic meteorologists saw cor-relations between the state of the upper atmosphere and surface weather conditions—high values of ozone with low-pressure systems, ionization with thunderstorms, and radio echoes with the movements of polar, maritime, and equatorial air masses.[105] There were no theoretical models, however, of how the vast changes in this attenuated region influenced weather at the surface.

Anne Louise Beck

"U.C. Woman Will Study in Norway" reads a headline in the *Berkeley Daily Gazette* for August 4, 1920. A similar notice appeared in the new *Bulletin of the American Meteorological Society*.[106] Thanks to a $1,000 competitive Mowinckel fellowship from the American-Scandinavian Foundation, Anne Louise Beck, a 1918 honors graduate of the University of California, was going to the new "weather college" in Bergen to study meteorology and oceanography for a year under professors Bjerknes and Helland-Hansen. Beck had work experience as an assistant in the Berkeley astronomy department, in the San Francisco office of the U.S. Weather Bureau, and as a high school mathematics instructor. She was enrolled in a master's degree program at Berkeley and was a star pupil in Burton M. Varney's geography course 121, Current Developments in Meteorology and Climatology, which emphasized readings, reports, and conferences on recent matters in these sciences. It was here she likely learned about Bjerknes's work and the fellowship program.

Beck set off on her adventure from New York on the Norwegian America liner *Bergensfjord* on August 27, 1920. She did not stay with the Bjerknes family. Advertisements running in the *Bergens Tidende* and several other papers announced that a "female American fellow to study this winter at the Geophysical Institute desires a centrally located furnished room, preferably with a full pension."[107] Her early impressions of the program appeared in the *Bulletin of the American Meteorological Society*:

The Bergen Geophysical Institute has two divisions. The A. division in charge of Professor Helland-Hansen carries on investigations in Oceanography both chemical and physical. We have a regular schedule of lectures Wednesday and Friday from 9.45 to 10.45 A.M. Later some laboratory practice is to be taken up supplemented by actual research and observations in the nearby fjords. These lectures I am to assist in editing for publication as a possible text in Oceanography.

Geophysical Institute B., the Meteorological Division, is in charge of Professor V. Bjerknes. A weather bureau, in connection with the Institute, forms a splendid laboratory. The staff of the weather bureau includes besides the Director, Mr. J. Bjerknes, now on a lecture tour to the southern countries of Europe, two Swedish meteorologists, Mr. E. Björkdal, Director Pro. Tem., Mr. C. G. Rossby, Mr. A. Tveten and a number of assistants. Mr. Tveten is especially concerned with the problem of the nucleus of condensation in raindrops and is daily making observations by pilot balloon for use in the forecasts and for the pilot of the passenger aeroplane from Bergen to Haugesund and Stavanger. Mr. Björkdal and Mr. Rossby are in charge of the weather forecasting.

A weather chart is prepared 3 times daily, at 8 A.M., at 2 P.M., and 7 P.M. There are three forecast districts, one for northern, one for western, and one for eastern Norway, the forecast centers being at Tromsö, Bergen and [K]ristiania, respectively.

An entirely different method of forecasting from that of the U.S. Weather Bureau is used by the Norwegian Service. Messrs. Solberg, J. Bjerknes, and Bergeron, at the Bergen Institute have developed the theory that the phenomena of the weather of the Northern Hemisphere are largely dependent upon the surface of junction of polar and equatorial air. This line of discontinuity can be detected at the earth's surface by conditions of temperature, pressure, hourly pressure change, wind direction and force, humidity and visibility. The line of discontinuity passes through the centers of cyclones connecting the center of one with those of the preceding and succeeding cyclones. The polar air at the surface is identified as being cold, dry, very transparent, usually blowing from [a westerly] point, while the air identified as equatorial is warm, moist, with poor visibility, and blows from a [southerly] point.

The polar front of each cyclone or surface of demarcation of these two air types is divided into a steering surface and a squall surface. The Bergen weather bureau associates most of the phenomena of cyclones with different parts of the polar front, and in particular on all synoptic charts set out definite rain areas in connection with the two surfaces which meet in the cyclonic center. The forecasts are built almost entirely on the movement of these surfaces.

No formal lectures have as yet been given in Meteorology at Geophysical Institute B., but Professor Bjerknes has promised to give some mathematical treatment of the problem in the very near future. Before Mr. J. Bjerknes left on his southern trip, several informal discussions of the Bergen Theory had been given.

So far you may be glad to note I have had no difficulties because I could not speak or understand the Norwegian language. Many people speak English fluently and all are willing to practice it whenever an opportunity presents itself.[108]

Beck returned to the United States on June 6, 1921, on the steamer *Bergensfjord*, her head full of Bergen school theory and her trunk stuffed with weather maps and charts. She stopped in Washington, DC, to visit Weather Bureau director Charles F. Marvin, who had been a referee the previous year for the fellowship competition. Marvin expressed great interest in her work and, at the end of their conversation, offered her a position at Weather Bureau headquarters. She regretfully declined his offer because it was too far from her home in California. After returning to Berkeley, Beck wrote to Professor and Mrs. Bjerknes informing them that she had been awarded a fellowship in astronomy and was teaching six classes each week but had not yet decided between astronomy and geography for her master's work.[109] Professor Varney advised her to write about the meteorological work in Bergen, which she did. Beck completed her MA in geography at Berkeley in 1922, submitting as a thesis "An Application of the Principles of Bjerknes' Dynamic Meteorology in a Study of Synoptic Weather Maps for the United States."[110]

Her article in the *Monthly Weather Review*, derived from the thesis, presented a summary of the dynamics of the circular vortex with applications to what Bjerknes had called the "planetary vortex" in Earth's atmosphere.[111] She had helped Bjerknes prepare the text of his 1921 paper, "On the Dynamics of the Circular Vortex with Applications to the Atmosphere and Atmospheric Vortex and Wave Motions," during her fellowship year, a paper Carl Ludvig Godske called "perhaps the most fundamental and also the most elegant and inspiring paper Bjerknes has ever written."[112] This was in no small measure due to Anne's editorial input, and she knew the theory quite well. She explained how general hydrodynamic considerations led to the practical notions of the polar front and to families of cyclones propagating along it. She also prepared a sketch of the general circulation that linked cyclone families with larger planetary flows (figure 2.6). She was well schooled in the history of ideas, giving proper credit to Dove's theory of the struggle between equatorial and polar currents, Helmholtz's work on the formation of a surface of discontinuity between cold polar air and warm equatorial air, and the cyclone models of Mohn, Margules, and Shaw.

Beck's references to colloquial terms are colorful and undoubtedly reflect the language used during informal discussions in the Geophysical Institute to describe new phenomena. For example, she refers to the cold circumpolar vortex as the "polar calotte" or cap, cold air pressing southward as a "spreading cold tongue," nascent waves on the polar front as spiral-formed "breakers," and the three vertical cells of the general circulation as "running like cogged wheels." Following Bergeron, she called the cyclone stage, when the cold front just overtakes the warm sector, a "seclusia" and the final stage, when the warm front is completely overtaken by the cold, an "occlusia." Summertime showery weather was referred to as "amoeba" cyclones. In discussing families of cyclones, the first wave of a new series was jestingly called "Protesilaos," the Homeric hero who returned from the dead. She recommended, as had Bjerknes, that in the future, wind directions should be recorded in more than eight compass points to help with the identification of fronts, that systematic cloud observations should be taken to document frontal passages, and most important, in agreement with Bjerknes, that some 4,500 telegraphic stations should be established across America to match the density of the ninety or so in Norway.

Beck prepared maps using Bergen techniques to analyze U.S. weather observations for the thirty-one days of January 1921. Her analysis for the first day of January appears in figure 2.7a. In the opinion of meteorologist Lou McNally, Beck's understanding and use of the Bergen school wave-cyclone model and the procedures for identifying the polar front on a synoptic surface map placed her "at the cutting-edge of forecast theory and application." Although she was confronted by vast regions with sparse data and stations with poor observations, the discrete steps Beck learned in Bergen for identifying fronts helped her overcome these lacunae, producing a viable analysis.[113] It was a completely different story when Beck sent the maps to the *Monthly Weather Review*. Only the map for January 1 was published and only after heavy-handed redrawing by the editor, A. J. Henry, rendering the original cold front unrecognizable and adding an unsupportable arrow indicating warm-air advection (figure 2.7b). Citing space limitations, Henry did not publish Beck's detailed discussion of the groups of cyclones that crossed the United States in January 1921. He suggested that interested readers could view them in the Weather Bureau library, where the originals "are available to students and others who may wish to consult them." They are not there; they were misplaced. Henry also criticized the idea that the United States could afford to establish and staff 4,500 weather stations; at the time they had only 200. He also believed that the Bureau could not possibly analyze the data from that many stations rapidly enough to issue timely forecasts. Weather Bureau resistance to Bergen school methods was fueled by these perceived limitations, by significant geographic differences between the two nations, by the entrenched Washington forecasting bureaucracy, and by the notion that Norway was in no position to tell the United States what to do.

Further details about Anne Louise Beck are hard to come by. She taught high school in Santa Rosa, California, for several years before joining the faculty of Santa Rosa Junior College in 1929, the year the college gained its

A

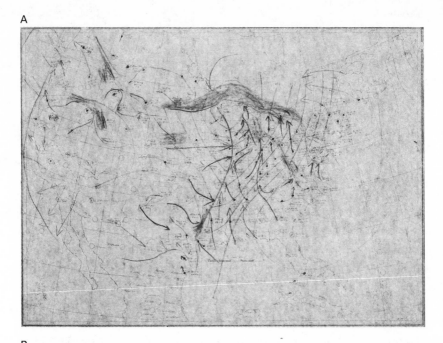

B

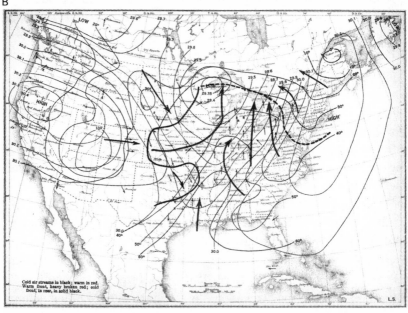

Figure 2.7
(a) Anne Louise Beck's original analysis for January 1, 1921, from her Berkeley thesis. Fronts, wind shifts, and shaded areas of precipitation are clearly marked. The original is drawn in white on blue drafting paper. Here the image has been inverted and the contrast enhanced. There are thirty-one maps in Beck's thesis, one for every day in January 1921. (b) Beck's map for the same date as published in the *Monthly Weather Review* but with heavy-handed manipulations by editor A. J. Henry.

independence from the high school.[114] She initially taught astronomy and mathematics and later geography. Her course on meteorology was described in the catalog as "Atmospheric changes that determine weather conditions; development of weather observations into climatologic data." The students received a thorough introduction to the subject. She taught courses in art, her hobby, and advised the campus honor society Alpha Gamma Sigma.

The Great Depression took a huge toll on the college. With no work available for high school graduates, enrollments swelled, but student motivation decreased, along with faculty salaries. As described by the dean of faculty: "The junior college instructors, faced with the task of teaching students who were discouraged at the outset and made little more than a half-hearted effort to learn, were hard pressed to pursue academic paths." Education at Santa Rosa became "primarily custodial in nature" as admission standards slipped and the original mission of the school, which had been to prepare students for more advanced study, was expanded to include vocational training and cultural programs for adult learners. Ironically, in the depths of the financial depression, the college acquired its forty-acre campus, the former experimental gardens of Luther Burbank, and inaugurated an aggressive building program.[115]

The 1930 U.S. Census lists Anne L. Beck (born 1896, single, age thirty-four) as living at 240 Carrillo Street in Santa Rosa. Her image in figure 2.8 is from the 1930 college yearbook, the *Patrin*.[116] The 1938 yearbook gives her married name as Anne Beck Walker. Emulating the interests of her mentor Bjerknes, she taught theory of aviation and meteorology in the Civil Aeronautics Authority program established in 1938, with the flight school conducted at the Santa Rosa Municipal Airport. After 1941, when the government banned private aviation on the Pacific coast, instruction was relocated to Ely, Nevada. The 1940 census lists Anne as living in Santa Rosa at the same address (married, head of household, age forty-four), with the annotation, "husband gone indefinitely, no further information." Her family—father Peter (age seventy-five), mother Margaret (sixty-six), and younger sister Margaret (forty-one)—still lived in Berkeley. The last mention of Anne on the Santa Rosa faculty is in the 1941 *Patrin*. She resigned from the college that year and, under contract with the U.S. Army Air Forces, joined the Cal-Aero Academy near Ontario, California, as an instructor in meteorology.[117] Here the trail goes cold. There is a death record for

Figure 2.8
Anne Louise Beck, 1930. Image courtesy of Santa Rosa Junior College Archives, Santa
Rosa, California.

Anne Beck Walker (born March 18, 1896, died November 29, 1982) in
Fresno, California.

Anne Louise Beck was two years older than Carl-Gustaf Rossby. She held
the same type of Scandinavian exchange fellowship (but six years earlier
and in the reverse sense). Beck and Rossby overlapped in Bergen in the
autumn of 1920. Both had the same level of educational attainment. Both
worked to bring Bergen methods to the United States. Beck wrote her MA
thesis on this topic in 1922 and published a long article that same year, so
her effort preceded Rossby's by several years. Her relationship with Weather
Bureau chief Marvin was cordial, so far as it went, while Rossby's was not.
Beck taught meteorology to pilots and Rossby established a model airway
weather reporting system in Oakland, California (chapter 3), only a short
distance from Beck's family home in Berkeley. Both Beck and Rossby worked
in World War II weather training programs, yet there is no evidence their
paths ever crossed. Beck's effort to bring Bergen methods to the U.S. Weather
Bureau in the early 1920s was a first step in a process that required a massive change in Weather Bureau culture and reached completion only
through the concerted efforts of many meteorologists including Rossby,
working over two decades (chapter 3). Her map of fronts and a cyclone,

published in *Monthly Weather Review*, was heavily manipulated and distorted by a prejudicial editor, A. J. Henry, who had been with the weather service since 1878 and whose perspective on the value of her work was further distorted by institutional and national pride, ageism, and probably sexism. Gender expectations of the time both shaped and severely constrained Beck's subsequent career path. Her story holds added significance because the most common dating for the emergence of prominent women scholars in meteorology is 1949, when Joanne Simpson received her PhD. Anne Beck was twenty-seven years her senior and made a significant intellectual contribution to meteorology at a time when this advanced degree was not required or even expected.

In Transit: Bjerknes's Second Transatlantic Trip

Vilhelm Bjerknes very much wanted to lecture on the new Bergen methods in Washington, DC and send one of his assistants to the U.S. Weather Bureau to carry out "practical reforms." When Charles F. Marvin heard of these evangelical plans from the new Carnegie Institution president John C. Merriam, he was "a good deal concerned." He informed Merriam that the U.S. Weather Bureau was "fully acquainted" with the theoretical writings of Bjerknes and his followers and had no need for their services because they already had "perfected" forecasts and warnings, both for the immediate needs of seafarers and for a week or more in advance for all. A flurry of retractions and cross-town consultations ensued.[118]

Vilhelm and Jacob Bjerknes, along with three hundred or so delegates bound for the British Association for the Advancement of Science meeting in Toronto, sailed from Liverpool on the *RMS Caronia* in late July 1924 and arrived in Quebec City in early August. Among the illuminati on board were Lords Ernest Rutherford and William Bragg and their wives. The meteorological cognoscenti included Richardson, Harold Jeffries, and Shaw. The Bjerknes's set up a small weather station on the deck during the voyage and entertained the passengers with daily forecasts.

In Toronto, the elder Bjerknes lectured on "The Forces Which Lift Aeroplanes." Reminiscent of what he had done with his own father in Paris in 1881, "Prof. Bjerknes and his son demonstrated with paper and wooden cylinders the application of their theory." It was a show-and-tell. Bjerknes used a ribbon to rotate and launch one of the paper cylinders out over the audience: "It rose gracefully in the air and soaring across the room looped the loop and almost landed in the lap of Sir William Bragg."[119] He had the audience move back before his son launched a heavy wooden cylinder in the same manner. He even repeated the old water tank experiments with rotating cylinders to demonstrate attractive and repulsive forces. Bjerknes then launched into a critique of Newtonian action-at-a-distance theories and referred to Faraday, Maxwell, Hertz, and his father, Carl Anton Bjerknes, in support of fluid fields of force that emulate this phenomenon. As he had done many times and so many years before, he explained the forces at work by comparing them to the interaction of electromagnets and electric fields:

At any moment the field of motion of the fluid system is identical in geometrical structure with a certain static or stationary magnetic or electric field; and in this hydrodynamic field there are *apparent* actions-at-a distance which, element for element, are equal but opposite to those of the corresponding forces of the magnetic or electric field.[120]

The key word here is "apparent" because he was presenting a field theory that opened up the possibility of applying electromagnetic formulas to the theory of airplanes. "Whirling Cylinders May Replace Wings on New Aeroplanes" read the headline in the Toronto *Globe*. Bjerknes ended his lecture by remarking, quizzically, that "the forces that lift aeroplanes may also have something to do with forces acting in atoms," but there is no record that he ventured any further into quantum theory.[121]

Jacob Bjerknes lectured on "The Importance of Atmospheric Discontinuities for Practical and Theoretical Weather Forecasting." He offered his audience a "mathematical treatment" of surfaces of discontinuity that would make the current qualitative rules into quantitative ones, earning "forceful applause," for his efforts.[122] The elder Bjerknes met Weather Bureau director Marvin at the meeting and most likely discussed how to reconstitute a complete circumpolar weather service; Jacob would have mentioned his upcoming research visit to Washington, DC. Vilhelm also met Robert Millikan in Toronto, who had invited him to present a course of lectures at Caltech on hydrodynamics and its applications to terrestrial—and possibly cosmical—physics. At the conclusion of the meeting Jacob headed east, and Vilhelm joined the British Association excursion to the Pacific coast. He wrote to Robert Woodward that he would later like to call at Washington, "as soon as a Northern European can stand the heat there."[123]

The delegates traveled across Canada in a Pullman car, stopping in Jasper, Alberta, in the Canadian Rockies; then on to Vancouver, British Columbia; and finally to Victoria, where they were feted at a reception and dinner hosted by the mayor and city council. Bjerknes's itinerary took him from Victoria to Blaine, Oregon; then to Berkeley, California, to visit mathematics professor M. W. Haskell for several days and present a lecture; then on to La Jolla to visit geologist T. Wayland Vaughan, the new director of Scripps Institution for Biological Research. He lectured to the "intelligence club of La Jolla," met with naval officers from San Diego, and watched the takeoff and arrival of aeroplanes.[124] In Pasadena, Bjerknes met physicist Chandrasekhara Venkata Raman and astronomer George Ellery Hale, toured Mount Wilson Observatory, attended a talk by astronomer Arthur Eddington, and prepared a lecture series on physical hydrodynamics and an article on solar hydrodynamics.

In his lectures, Bjerknes reviewed the basics of vector notation, statics, and kinematics, then presented his views on the forecasting problem and the history of meteorology:

In the year 1875 the immortal Helmholtz gave a popular lecture, "Wirbelstürme und Gewitter." He begins by quoting a verse; he had forgotten who is the author . . . but he had never been able to forget the verse itself:

Es regnet, wenn es regnen will
Und regnet seinen Lauf
Und wenn's genug geregnet hat
So hört es wieder auf.
—Carl Friedrich Zelter (alt)

It rains when it wants to
It rains all it wants to
And when it is all rained out,
It stops.

The reason is obvious, he says: its content touches a sore spot in the conscience of the physicist:

Under the same firmament on which the stars move as the symbols of invariable laws of nature, the clouds form, the rain pours, the winds change apparently as symbols of the opposite principle. Amongst the phenomena of nature they are the most capriciously variable ones, the most fugitive, the most impossible to grasp; they escape every attempt of ours to catch them in the enclosure of law.[125]

For Helmholtz, the mutable clouds and the crashing rain, operating under the same vault of heaven as the fixed and orderly stars, appeared as opposite extremes, the weather seemingly having escaped the constraints of natural law. Yet he never stopped trying to explain the natural order. In 1888, while vacationing in a mountain resort, Helmholtz interpreted the dynamics of altocumulus "billow-clouds," explaining them by the gravitational waves that still bear his name. He also had a rather vague notion of a world-encircling atmospheric surface of discontinuity caused by the air motions of the general circulation. However, he could not think systematically about the bigger systems of motion because he lacked the necessary worldwide observations.

Bjerknes made it clear that wider and higher observations were desirable. Technologies such as the telegraph and the Atlantic cable had made storm warning quicker and more widespread, but the formulation of empirical rules was no substitute for physical understanding of the laws of nature. In the late nineteenth century, with the principles governing atmospheric changes not forthcoming, the compilation of statistics achieved prominence, and climatology became a dominant practice. Aerology held great promise, but the fliers of kites and balloons were empiricists too and had yet to identify governing principles.

Outside the realm of meteorology Euler, Lagrange, Claude-Louis Navier, and George Gabriel Stokes had provided the mathematical foundations of hydrodynamics; Sadi Carnot, James Joule, and William Thomson, thermodynamics; and Gaspard-Gustave Coriolis, relative motion on the rotating earth. C.H.D. Buys-Ballot had an intuitive grasp of how this related to winds, and William Ferrel, having rediscovered the Coriolis principle, developed a theory of the general circulation and of the direct thermal forcing of cyclonic and anticyclonic vortices. In Norway, Cato Maximilian Guldberg and Henrik Mohn reinforced this view, which remained dominant until Julius Hann discovered the astonishing fact that ascending air in cyclones was in general colder than its environment, meaning their energy came from some more fundamental process. The theories of Dove, involving the clash of polar and equatorial currents, inspired Wilhelm Heinrich Blasius, Helmholtz, and Margules to take up anew the dynamics of cyclone formation on a larger scale.

Bjerknes ended his lecture as he had begun, by quoting Helmholtz, who was in essence paraphrasing Laplace:

[Our confusion] is only the expression of the incompleteness of our knowledge, and the dullness of our power of combination. A mind which had the precise knowledge of the facts, and whose mental processes would work rapidly enough to hasten on in advance of the events, would recognize in the wildest capriciousness of the weather no less than in the motions of the stars, the harmonic rules of eternal laws, which we now merely presuppose and guess.[126]

Precise knowledge of the facts might come from a global observation system, and the pace of computing would certainly hasten on in advance of events, but it is the informed mind and the theoretical model that makes the system work.[127]

In mid-December, Bjerknes left Pasadena and wrote to his wife from a train heading east:

Dear Nora. Now it is done. I roll eastwards through the desert. The Pasadena adventure belongs to the past. Everything considered, it went quite well. The competition from the greatest physicist from Asia [Raman] has not been easy. The turban and the oriental eloquence have been good cards in his hand, especially where the sensational plays such a huge role as here. And not being a modern atom physicist has been strongly to my disadvantage. I have been the fifth wheel on Millikan's triumphal car.[128]

He also described the "immense amount of things that make life more easy and comfortable" in America and said that he had almost purchased a car. The rest of his itinerary included a lecture in Minnesota, Chicago for Christmas, and then on to Washington, DC, where he delivered two semipopular lectures on January 6 and 7, 1925. The first, "On the Forces That Lift Airplanes" (with experiments), was a reprise of his Toronto lecture. The second, "Solved and Unsolved Problems in Dynamical Meteorology" (with projections), provided an exposition of what had been achieved in meteorology with support from Carnegie and what still remained to be done. This involved applying the laws of hydrodynamics and thermodynamics to atmospheric and oceanic motions, giving special emphasis to the present state of the "polar-front" theory of cyclones and anticyclones and the general atmospheric circulation.[129] The lectures were presented not at the Weather Bureau but in the 200-seat auditorium of the Carnegie administration building, with special invitations issued to members of the Washington Academy of Sciences. He also spoke at Columbia University before departing on RMS Antonia for England. He wrote to Honoria, "A lovely day. We have Jack's polar air with sunshine and cumulus clouds."[130]

Accomplishments and Setbacks

We can predict the weather by calculating *new positions* of the moving air masses and the knowledge of their *physical states* in these new positions, a problem as you can't imagine being solved except by the method of Lagrange.
—Vilhelm Bjerknes, 1933[131]

The five-month trip allowed Vilhelm Bjerknes time to reflect on his intellectual trajectory. When he first received Carnegie support in 1906, he planned to limit himself strictly to the working out of formal methods, leaving it to the meteorologists to make practical use of them, but as the work proceeded, he realized he needed a deeper empirical knowledge of atmospheric conditions, especially if he hoped to reform any of the existing practices of the weather services. In Leipzig, he conducted empirical work in parallel to the theoretical investigations, and in 1918, driven by pressing needs, he organized for the Norwegian government a weather service for western Norway, the *Vervarslinga på Vestlandet*. His previous Carnegie assistants adopted a new point of view and used new methods to make daily forecasts. This, in turn, produced new empirical results of both immediate practical value and high theoretical importance. Discontinuities, which previously had been regarded as exceptional phenomena, now came to be seen as actually governing the weather.

Inspired by Bjerknes's program, Lewis Fry Richardson attempted a strategy of treating meteorological phenomena mathematically in all their complexity—head on—through the laborious method of mechanical integration. It didn't work.[132] He used data for central Europe from one of the international observing days (May 20, 1910) and worked for eleven years (1911–1922) to construct a six-hour forecast using the complete hydrodynamic equations. He was metaphorically just beginning to chisel a tunnel through stone, however, since the observations he used were sparse, the computational scheme using unfiltered equations was unstable and intractable, and his result, a calculated 40-millibar pressure rise, was unrealistic. Another unintended result of Richardson's unsuccessful trial, according to a later account by Bergeron, was to put a damper on others who were thinking of attempting their own computations. Richardson's forecast can then be seen as a quixotic numerical experiment conducted with manual labor, a futile attempt to untie the forecasting knot for one particular day. He demonstrated what might have been predicted—failure. Richardson was not the "first" to try this, as some have claimed (Exner and Bjerknes attempted limited numerical computations years earlier), nor did Richardson exert a direct influence on the development of computerized numerical forecasting (he left the field of meteorology in the 1920s, feeling that it was too militarized for his pacifist ideals).

Bjerknes wanted to work on an important part of the problem, not all of it. He realized that cyclonic disturbances propagating along an atmospheric surface of discontinuity could be treated as wave motion and could be reduced to solvable linear differential equations. The full solution still required a great amount of mathematical work guided by the latest empirical findings. With the Caltech lectures fresh in his mind, Bjerknes was determined to make a serious attack on this problem and expressed Newtonian confidence that its solution "will prove to be of similar importance for meteorology as that of the motion of two attracting bodies has been for astronomy." He wanted to rehire his ablest collaborators who had moved on to better positions, so he asked Carnegie for $5,400 per year for three years, a threefold increase over previous requests. He promised to reedit the earlier volumes of *Statics* and *Kinematics* and issue them simultaneously with the long awaited third volume, *Dynamics*, "which should contain as its concluding chapter the solution or at least [as thorough a discussion as possible] of the defined problem of the cyclones."[133]

Bjerknes was thinking globally and even cosmically. In 1924, he reported to Merriam that his quest for a completely circumpolar weather service had now "become a reality," with pressure maps being drawn, at least on favorable days, for the entire polar region and a great part of the Northern Hemisphere. He wanted to bring the meteorological phenomena on Earth into connection with their "ultimate cause," solar radiation, and asked Hale's opinion about incorporating measured values of the solar constant into the daily meteorological analysis. He was also thinking vertically because the Norwegian weather services had recently instituted an aviation program to improve both forecasting and flight safety. Ernst Calwagen, manager of the Bergen meteorological observatory and a forecaster for the military air forces, brought life and excitement to the program. He flew in instrumented aircraft, photographed cloud systems, and coordinated his observations with ongoing analysis on the ground. Tragically, Calwagen and his pilot died in an airplane crash in August 1925, bringing this promising program to an end. Twenty years later, Bjerknes recalled Calwagen's unforgettable utterance: "To analyze a weather map after Bergen methods is like reading the most exciting novel."[134]

Bjerknes, now working in Oslo, sent his annual report for 1928 to the Carnegie Institution in late August. It had been a year of setbacks. His wife, Honoria, had fallen seriously ill and died in the spring, and his close assistant, Halvor Solberg, had been bedridden for several months and was still weak. Although these blows interfered seriously with his work, he found consolation in pressing on and wrote optimistically about his prospects for completing a paper on the Lagrangian equations of motion. It was a long report in which Bjerknes looked back on twenty years of support and used the occasion to itemize the serious risks he had taken. Here is part of the report in his own words:

I took a risk when in the year 1906 I made my first proposal to the late president Woodward of taking up this work. I knew that my proposal was a rational one. But it was not possible at that time to foresee to which degree Mr. Sandström's and my work would meet the potential needs of meteorology at that epoch.

I took a risk in the year 1913, when I ventured to go to the foundation of the Geophysical Institute of the University of Leipzig, with the program to use the two first volumes of my Carnegie work for working out synoptically in three dimensions the international aerological observations then organized in Europe, hoping to find out what actually happened in the three dimensional atmosphere. But the war broke out and interrupted our work before we had had opportunity to state if the risk had been justified.

I took a risk when, in the year 1918, I induced the Norwegian government to establish a weather service at Bergen, to try if the methods for weather forecasting developing from my Carnegie work would not succeed even without the help of the direct aeorology which we had had at our disposal in Germany. This risk of mine gave my young collaborators, J. Bjerknes, H. Solberg, T. Bergeron the opportunity to discover the polar front, and to develop the methods of "indirect aerology."

I took a risk also in the year 1925, when I asked for the increased grant for a last attack upon the problem of the cyclones, simultaneously from the empirical and the mathematical side. I had a strong personal confidence to be on the right track with my "wave theory of cyclone formation." But I knew that perhaps most competent meteorologists thought that I was wrong, and I could not hide for myself that the problem might exceed our forces. But my equations of atmospheric disturbances and Mr. Solberg's integration of them have fully justified even this risk.[135]

Bjerknes was sixty-six years old and ended his report with a risk of another kind—asking for a continuation of his grant. He still was in good health and still had full mental forces, but he wanted to retire, not from work but from the demand of support, to avoid the risk of being informed by others that he was no longer the "right man." He was not going to be able to complete the final volumes of *Dynamic Meteorology and Hydrography*, but he hoped that the Carnegie Institution would "take the revolution, which has been produced in the science of meteorology, as an equivalent of that work."[136]

Bjerknes returned to Paris in 1929, almost four decades after his first visit, to present twenty-nine lectures on physical hydrodynamics similar to those he had given at Caltech in 1924. The goal was to provide hydrodynamics, the theory of the motion of liquid media, with a central position among the physical sciences. Bjerknes's analysis extended beyond the purely classical hydrodynamic treatment of ideal fluids, where established vortex conservation laws are valid, but the physical laws of thermodynamics do not apply. He was interested in explaining, mathematically, the real fluid behaviors of geophysical and cosmic systems. The volumes were published in German in 1933 and in French in 1934 with sections on dynamic meteorology provided by Jacob Bjerknes, Solberg, and Bergeron; they were dedicated "to the memory of C.-A. Bjerknes whose work has found a continuation in this work." In the foreword, Vilhelm Bjerknes indulged in soaring rhetoric on cosmic fluids and went so far as to invoke the old ether theory:

We humans live, breathe, and move in a gas-liquid medium, the atmosphere surrounding the Earth. Also, the sea that covers three-quarters of the Earth plays an important role in our lives, and a few kilometers below the surface, according to our present view, the solid Earth turns into a liquid state. The sun and the fixed stars that send us their light are gaseous-liquid bodies, and all the heavenly bodies appear to emerge from gaseous nebulae. And, if there is after all an ether filling space and transmitting light and remote effects, so must this ether share a common property with liquids, namely the property of allowing bodies to move through it. With Heraclitus we can say: πάντα ρεῖ, everything flows.[137]

The text itself is scientifically sound, presenting a clear distinction between the Eulerian and Lagrangian mathematical approaches, a summary of dynamic meteorology, and a historical, largely autobiographical perspective on the work.[138]

Although his official retirement came in 1932, Bjerknes remained the leader of the Oslo group for another five years (figure 2.9). He was at the

Figure 2.9
Vilhelm Bjerknes and team in 1934. Front row, left to right: Tor Bergeron, Vilhelm Bjerknes, Jacob Bjerknes, Carl Godske. Back row, left to right: Halvor Solberg, Harald Sverdrup, Sverre Petterssen. *Source:* Bergen Geophysical Institute.

peak of his influence, and honors flooded in. He was elected an honorary member of the German Meteorology Society and a foreign member of the Royal Society of London and the U.S. National Academy of Sciences; he received the Agassiz Medal for Oceanography, the Symonds Medal of the Royal Meteorological Society and the Buys Ballot Medal for Meteorology, awarded once each decade, from the Royal Dutch Academy of Sciences.[139] He was very pleased in 1934 to hear that the millibar, a unit of pressure he had proposed in 1912, had finally been approved.

When his Carnegie grant was renewed in 1935, Bjerknes, age seventy-three, wrote a "quite confidential" letter to Merriam describing the characteristics of each member of his staff and the ways they could continue the work in his absence:

> *Dr. H. U. Sverdrup*, professor, Bergen, 46 years of age. He was my Carnegie assistant 1908–16, in which year he joined Amundsen as scientific leader of his polar expedition then in preparation. His later career will be known to you: he passed 7 years in the polar ice, and has been occupied later mainly with

the working out the scientific results of the expedition. I believe you know him personally, and I think you will agree with me that he is a man of great intellectual and moral value. He has avoided the great danger for scientists of expeditions, to become superficial. He was absent from my work at the time of its greatest progress, which was connected with the foundation or the Meteorological Central at Bergen on the West Coast of Norway 1918. He hardly masters the "art" of "Weather Map Analysis" and of "Weather Forecasting" in their most refined form, but he is familiar with the general principles. He is without comparison our best expert both in practical and scientific oceanography.

Dr. J. Bjerknes, professor, Bergen, my son, 35 years of age. He has been attached to my work without interruption since the year 1915, first as one of my Carnegie Assistants, and then as the actual leader of the new Meteorological Central at Bergen (I was merely formal leader, and that only during the first years). What is generally known as the "Bergen Methods" or the "Norwegian Methods" of Weather Map analysis and of Weather Forecasting is mainly due to him conjointly with Dr. Bergeron. He may have more general experience in the entire field of our researches than anyone else of my staff, and in addition to his scientific value he has proved to be a good organizer and administrator. You may have seen him during his visit to U.S. and to Canada last year. He enjoys great confidence and personal popularity among his fellow meteorologists in the different countries. Just in these days he has organized the sending up of a great number of registering balloons from a great number of European states during the passage of a big extra-tropical cyclone.

Dr. H. Solberg, professor of theoretical meteorology at the University of Oslo, 37 years of age. He became my Carnegie Assistant in 1916, and remained, with some interruptions, until a few years ago he became professor. He is a mathematician of very high value, and I have often in my annual reports given expression to my admiration for his achievements. He is not interested in organizing or administrative work; he is always absolutely absorbed by his problems.

Dr. T. Bergeron, meteorologist, Oslo, about 40 years of age, has with some interruptions been attached to my work since the year 1919. Dr. B. is a most remarkable personality: high intelligence, rich in ideas, idealist, enthusiast, great capacity of work. He has made us invaluable services, although it is very difficult to make him finish a work, or to make him reduce a paper to its simplest possible form. But no one is comparable to him as a personal instructor in the difficult art of Weather Map Analysis, of which he is one of the founders. He has several times been invited to foreign Meteorological Institutes as instructor in the "Norwegian Methods," and has always had great progress. He has passed half a year in English service at the island of Malta (meteorological service for the air route England-India), and he has twice been in Russia in the service of the Soviet, the last time more than half a year. He learns with ease the most difficult foreign languages. The best way of using his rare faculties is to give him full freedom to follow his own ways, and not to prescribe him anything. If the "Norwegian Methods" should be really introduced in the U.S., no one would be a better teacher in these methods than he.

Dr. C. L. Godske, about 27 years of age, Oslo. He is a brilliant mathematician and an exceedingly energetic worker. His interests are of a similar universal nature as those of Bergeron, but at the same time he is very different from him: he is a very systematic worker who always finishes one task before he takes up the next. He is at present the only one who is paid in full by the grant, and devoting himself entirely to the work, he makes us great services by his great capacity of work. I have every reason to believe that he will become an important man of science.

Dr. S. Petterssen, leader of the Meteorological Central, Bergen (as successor of my son), middle of the thirties. He has introduced valuable novelties in the practical methods of weather forecasting; this made me invite him to join our Carnegie work. He is a reliable systematic worker, and a good organizer and administrator. He has not had the same chance as the others to begin his scientific work early, and thus his scientific basis is somewhat narrow. But he seems to develop well in every respect. You may have the opportunity to see him, as he has been invited to come to Washington this year. I believe he will come in the month of May. [See chapter 3; he was a guest of U.S. Weather Bureau chief Gregg.][140]

Then Bjerknes got to the point. Although feeling well both mentally and physically, he recommended either Sverdrup or Jacob Bjerknes as his successor. To confirm his vitality, he enclosed a photograph of himself on a mountain expedition where the temperatures averaged minus 27 degrees centigrade.

Bjerknes was pleased with the inroads the Bergen school was making in England. In 1935, Ernest Gold published his fifty-page presidential address to the Royal Meteorological Society on "Fronts and Occlusions," complete with diagrams provided by Jacob Bjerknes. For the silver jubilee of King George V, George Clarke Simpson wrote a prominent article on "Weather Forecasting" devoted to the Bergen school, concluding that weather forecasting was developing from an art to a science. Sir Napier Shaw, the grand old man of British meteorology, devoted the final chapter of his popular book, *The Drama of Weather*, to Bergen school methods. At the time, Jacob was in London serving as an instructor for five months in the Meteorological Office on how to prepare forecasts for transatlantic flights using circumpolar charts and an extended system of weather telegrams from ships.[141]

In 1937, Bjerknes became a member of the Pontifical Academy of Sciences, a scientific society that traced its roots to the 1603 Accademia dei Lincei. The society, recently re-constituted by Pope Pius XI, was dedicated to scientific research beyond the influence of national, political, or religious factors. Nevertheless, the new inductees were worried about the state of science in Germany as indicated by the Berlin Academy's recent change of statutes to align science with the Nazi regime. Austria and Italy had made similar moves in the universities where the deans were appointed by the government and in restrictions on travel and exchange of information with foreign colleagues. Bjerknes's 1937 visit to Berlin illustrated this:

Everything is controlled. I telephoned in Berlin to the Telefunken Company (a private manufacturing company for wireless material) saying that I wished to call on old scientific colleagues working there. I was told that I would be welcome, but I had first to get the permission of the government, which I would certainly obtain. I had not time to wait for it. When I visited Reichsamt für Wetterdienst (the Government Weather Service) (which works now with enormous material resources, and uses the Norwegian methods), it was noted carefully the hour and minute when I entered and when I left, and who had conveyed me. The notes must show that no visitor has been a single moment alone, and thus has had the opportunity to pick up secrets. The result is that a German scientist feels like in prison. But at the same time the press tells the people every day that Germany is the only really free country in the world, and as never an opposition is heard, a great part of the people actually believes it. That gives government its strength.[142]

Further details in Bjerknes's letter paint a grim picture of science under totalitarian rule.

In 1938, Bjerknes spoke in Leipzig at the twenty-fifth anniversary of the Geophysical Institute. After recounting a bit of history, including his own tenure from 1913 to 1917 and his subsequent work in Bergen, he added diplomatically, "One school is a direct continuation of the other, they only took different external forms." Bjerknes expressed his hopes for international cooperation on aerology between Bergen and Leipzig under the leadership of the current institute's director Ludwig Weickmann, who also served as head of the Third Reich's weather service and scientific adviser to the air ministry. Such cooperation was not to be. As it was in 1913, so too in 1938, one year before the outbreak of world war.[143]

In 1938, the Carnegie Institution installed a new president—engineer and inventor Vannevar Bush, who had been the vice president of MIT and dean of the MIT School of Engineering. Bush tried to get up to speed on

Bjerknes's long-promised third volume of *Dynamic Meteorology and Hydrography*. Bush asked his assistant Walter Gilbert, "Are we involved with the publication of this when complete?" Gilbert replied, "Very much so," explaining how Carnegie published volume 1, *Statics*, in 1910 and volume 2, *Kinematics*, in 1911 and had contracted for volume 3. The grant money had been flowing annually since 1906, but there was as yet no final publication, and Vilhelm Bjerknes had dropped off the writing team, leaving that to Jacob Bjerknes, Bergeron, and Godske.[144]

Bush sought Rossby's advice, asking him what, if anything, should now be done regarding this "very indefinite situation."[145] Rossby did not mince words:

I have for some time been of the opinion that this book should have been written either a few years ago or else that the writing should be delayed perhaps three or four years. This opinion is based on the fact that many new ideas have been developed here in the United States during the last few years concerning the general circulation of the atmosphere, partly as a result of the exceptionally fine network of upper air stations which we now have in operation over this continent, and partly as a result of the interpretative work that has been going on [at MIT]. . . . unless it takes these new results and ideas into account, [this volume] must necessarily become obsolete within a very short time.[146]

If Carnegie support were to continue, Rossby recommended that the writing team, with Jacob Bjerknes taking the lead, should "incorporate modern American research results into the treatise," including the latest results on atmospheric radiation.[147]

World War II was not kind to Norway or to Bjerknes. On April 9, 1940, the German army seized the weather prediction centers in Oslo and Bergen, and the Norwegian weather alert service was commanded to stop; many of its members worked underground for the resistance.[148] Bjerknes was stuck in Oslo for the duration, unable to communicate freely with Godske in Bergen, Bergeron in Stockholm, or his son Jacob, who had emigrated to the United States. Jacob found it impossible to get his father out of Norway. As he wrote to Bush, "Unfortunately the only ship connections between Norway and the US, through Italy, are now also disrupted, so the realization of the transfer must of course be postponed. The long way through Russia, Siberia and across the Pacific seems to be too problematic."[149] Jacob also found it impossible to work on the manuscript due to the censorship of his mail and concluded that "the only thing we can do is to wait."[150]

At war's end, Jacob Bjerknes was relieved to find his father "in good health and scientifically active." A generation earler, Vilhelm had helped his father, so too Jacob helped his. He returned to Norway for a sabbatical and took responsibility for late additions and modifications in the manuscript. Godske had become the chief editor of the final volume. The manuscript was "almost complete," but by their own admission, it was out of date. Vilhelm's preface to the volume acknowledged that the "frequent disruptions of the planned continuous teamwork" had "no doubt left their marks in the manuscript." He wrote that this book, like all others, will be superseded, but expressed hope that it contained "building material for the final edifice of the science of meteorology."[151] This was written in 1949. The Carnegie Institution received the manuscript in 1951, and the book was copublished by the American Meteorological Society and the Carnegie Institution in 1957. As Rossby had predicted, the 800-page tome was not widely read. The content of the book was almost a decade old, and the unfortunate delay in printing occurred at a time when theoretical meteorology was developing rapidly through numerical methods, worldwide aircraft observations, and new work on severe storms and precipitation formation. Thus, rather than a work at the cutting edge, it served as a reminder of past accomplishments. J. S. Sawyer's review in the *Quarterly Journal of the Royal Meteorological Society* emphasized the book's historical value concerning "the state of the science at the end of World War II," but the reviewer was clearly disappointed by the lack of coherence in the section on weather forecasting, which should have been one of the book's strengths. Sawyer concluded that the volume represented "the culmination of an epoch in scientific meteorology."[152] Vilhelm Bjerknes remained scientifically active in his later years. In 1943, he wrote a personal review of the Bergen school's history for their twenty-fifth anniversary festschrift, he received his final three-year grant from Carnegie in 1945, and he published his final scientific paper in 1949 on "dynamics and electromagnetism," a hearkening back to his youthful days in Paris with his father. He died in Oslo, Norway, on April 9, 1951, at the age of eighty-nine.[153]

According to Vilhelm Bjerknes, the central problem of the science of meteorology is weather prediction by rational dynamical-physical methods. He said this in 1902 in the *Meteorologische Zeitschrift*: "The goal is to predict the dynamic and physical condition of the atmosphere at a later time, if at an earlier given time, this condition is well known." He reiterated this in 1904 when he stated the problem of weather forecasting as an initial-value problem in mathematics involving the ideal gas law, the first law of thermodynamics, the conservation of mass, and the dynamical equations of an ideal compressible gas. Because exact solutions to this system of equations were impossible, he promoted the use of graphical methods.[154]

Bjerknes insisted on CGS units and worked to rationalize the analysis of observations. He accepted a call to Leipzig in 1913 to establish a geophysical institute. This gave him access to a dense network of upper-air observations. The tragedy and hardships of World War I made conditions insufferable in Germany, and in 1917 Bjerknes moved to Bergen, Norway, with his family and two research assistants, his son, and Solberg; Bergeron joined them later. This group established a network of stations sufficiently dense to identify cold fronts and warm fronts as components of an ideal wave cyclone model. Working in concert, the Bergen school delineated the properties of air masses and described the life history of cyclones generated along the polar front. Bjerknes once remarked, "During 50 years meteorologists all over the world had looked at weather maps without discovering their most important features. I only gave the right kind of maps to the right young men, and they soon discovered the wrinkles in the face of Weather."[155]

The Bergen school of meteorology was evangelistic, with Bjerknes hosting numerous visitors in Bergen, lecturing in foreign capitals, and sending acolytes, including notably, Bergeron and Petterssen, to international outposts. Bjerknes lectured in America in 1905 and again in 1924–25; he received research support for forty-two years from the Carnegie Institution of Washington. Among those he influenced, Anne Louise Beck spent a year in Bergen in 1920–21 and then worked to bring Bergen school methods to the United States; Rossby came to America in 1926, where he built new schools of dynamic meteorology at MIT and Chicago; and Jacob Bjerknes established the UCLA department of meteorology in 1940. Vilhelm Bjerknes's work on geophysical hydrodynamics established the scientific basis of modern meteorology, and the Bergen school methods provided effective tools for weather analysis and forecasting. Both in his own work and through his extensive influence on the field, Bjerknes established connections between theory and practice and between the geophysical and technological communities.[156]

Rossby remembered Vilhelm Bjerknes as "a man with a bushel of hair, a remote interest in his students, and a frugal way with his family." On a kinder and more profound note, one of his final collaborators, Godske, described the life work of Bjerknes as the creation of a new branch of science, physical hydrodynamics; the introduction of the circulation theorems; and the application of this science to dynamical meteorology and physical oceanography. He remembered Bjerknes as the incomparable team leader, the enthusiastic and inspired stimulator, the brilliant lecturer, and the good and kind man.[157]

3 Rossby

The principal task of any meteorological institution of education and research must be to bridge the gap between the mathematician and practical man, that is, to make the weather man realize the value of a modest theoretical education, and to induce the theoretical man to take an occasional glance at the weather map.
—Carl-Gustaf Rossby, 1934[1]

In 1956, *Time* magazine featured Carl-Gustaf Rossby on its cover and profiled him within its pages, writing, "The history of modern meteorology is inescapably paralleled by Rossby's career." The article added: "One man who did as much as anyone to raise meteorology to its present high estate is a likable, high-spirited, round-faced Swede named Carl-Gustaf Arvid Rossby. Most leaders of modern meteorology are friends or past pupils of Dr. Rossby's. The 'Rossby parameter' is important in up-to-date forecasting, and the grandest movements of the atmosphere are called the 'Rossby waves.'"[2] These statements are basically true—most active atmospheric scientists trace their academic lineage back to Rossby—yet there is much more to learn about him.[3]

Rossby was a theorist, system builder, world traveler, and bon vivant promoting international understanding and cooperation. His field of action included the *Vervarslinga på Vestlandet*, the Swedish Meteorological and Hydrological Institute, the U.S. Weather Bureau, the Guggenheim Fund for Aviation, Massachusetts Institute of Technology, Woods Hole Oceanographic Institution, the University Meteorology Committee, the University of Chicago, the American Meteorological Society, the Institute for Advanced Study, the International Meteorological Institute, and the National Academy of Sciences. Rossby got his start under Vilhelm Bjerknes and the Bergen school but soon set his own course. He developed his own school of thought, bringing to the study of meteorology and dynamic oceanography novel analysis techniques and influential theories about the behavior of upper-air winds and ocean currents. He benefitted from and further developed airplane and radio-balloon soundings, rotating tank experiments, and other technologies. He served his adopted country well, supervising the instruction of thousands of weather cadets during World War II. He had meaningful, if often somewhat tense, interactions with three U.S. Weather Bureau directors—Charles F. Marvin, Willis R. Gregg, and Francis W. Reichelderfer. He was a mentor and role model to those who studied under him and an inspiration and unforgettable colleague to his peers. His best students included Bert Bolin, Horace Byers, and Harry Wexler, and his influence reached everyone in the field, including Jacob Bjerknes, Jule Charney, Sverre Petterssen, and John von Neumann, to name but a few.

Young Rossby

Rossby was born on December 28, 1898, in Stockholm, Sweden, the first of five children of Arvid Rossby, a construction engineer, and Alma Charlotta (Marelius). By all accounts, he was spirited, quick-witted, and excitable, with diverse interests in music, geology, botany, and especially the orchids that grew in abundance on the island of Gotland, his mother's ancestral home. He was an excellent and conscientious student who studied at Stockholm's Högskola, passing his examinations in both Latin and natural science and earning the degree of Filosofie Kandidat in 1918, with mathematics, mechanics, and astronomy as his major subjects. During a year of additional studies in Stockholm, Rossby attended a lecture by Vilhelm Bjerknes on moving discontinuities in the atmosphere. Here he met the great man, and on the recommendation of Ivar Bendixson, his mathematics professor, Rossby applied for a position as a research assistant under Bjerknes. He said he wanted to spend a summer vacation in Norway anyway.[4]

Rossby joined the Bergen school on June 20, 1919, and left on December 15, 1920. He spent the first year as a Carnegie research assistant working with Erik Bjørkdal and Tor Bergeron before accepting a position as a staff meteorologist. He arrived as an inquisitive twenty-year-old student with no prior knowledge of meteorology, and he readily mastered the key theoretical aspects of dynamic meteorology as formulated by Vilhelm Bjerknes and the practical techniques then under development. He was full of helpful practical suggestions, such as revising the color scheme for fronts on weather maps, launching experimental sounding balloons, and organizing teams of forecasters to extend the summer experiment of 1919 into the autumn and winter, a volunteer effort that allowed for the analysis and forecasting of winter storms. His strong suits involved identifying fundamental issues, generating new theoretical ideas, and mobilizing people around him. He was organized and persuasive, a skillful weather map analyst. Bergeron called him a natural orator and a natural leader, "whose far-reaching ideas and high-flying plans often took our breath away."[5]

In 1921, Rossby accompanied Vilhelm Bjerknes to Leipzig for a year of advanced study at the Geophysical Institute and gained experience launching instrumented kites and balloons at the Prussian Aeronautisches Observatorium in Lindenberg. He was seeking connections between theory and observations of the upper air. In his first scientific publication, Rossby proposed a farsighted if somewhat impractical program to establish a network of aerological stations around the Norwegian Sea and as far northwest as Greenland.[6] At the time, synoptic surface maps were typically of limited geographical extent, and observations of the upper-air flow patterns were missing almost entirely. Thus, even the most proficient forecasters from the Bergen school focused their attention on the motion and behavior of small-scale waves along the polar front. German meteorologists were the exception. Employing W. J. van Bebber's concept of *Großwetterlage* (large-view weather),[7] they identified five typical storm tracks over Europe and entertained the possibility that weather systems of quite different dimensions can exist at the same time in the atmosphere, the smaller ones being steered by the larger ones.

Rossby spent the summer of 1922 in Bergen and then returned to Stockholm, where he worked for three years as a junior meteorologist at the Swedish Meteorological and Hydrological Institute under the direction of Johan W. Sandström. At the same time, he continued his studies in mathematical physics with the celebrated mathematician Erik Ivar Fredholm, earning the Filosofie Licenciat degree in 1925. During the course of his studies, Rossby served as the meteorologist on board several scientific cruises. In 1923, he sailed to Greenland on the *Conrad Holmboe* to establish weather stations for the Bergen school and to collect oceanographic and meteorological data under the scientific direction of the Swedish meteorologist Oscar Edlund. The ship flirted with disaster when it was caught in pack ice near East Greenland for six weeks and suffered such heavy damage that it eventually had to be scuttled.[8] In 1924, the Swedish Weather Service assigned Rossby to the Royal Swedish training ship *af Chapman* on its voyage round the British Isles. His duties included preparing weather forecasts and taking upper-air observations. Some have said Rossby never took his own observations, but he certainly did so on this voyage. He collected and decoded weather information sent by official and amateur observers via wireless radio, a process he found to be "hopelessly confusing." As a supplement to this technique, he used his own observations of wind, cloud formations, and differences in temperature between the air and the surface water to prepare forecasts according to Bergen school methods. He found this single-point technique quite acceptable, especially over the open ocean. To investigate the winds aloft, Rossby launched pilot balloons and tracked them with a special theodolite, a tripod-mounted telescope for the precise measurement of angles, stabilized to function properly on the rolling deck of the ship.[9] By charting the results of nine ascents, he was able to estimate surface wind stresses and calculate the dissipation and vertical transport of energy in the atmosphere at different heights above the sea surface. On August 10–11, 1924, with the ship pounded by heavy weather and veering dangerously close to the Irish coast, a seriously seasick Rossby issued a critical—and correct—forecast for turning winds that would put them back on course. Seeking more tranquil climes, Rossby set sail once again in 1925 as the meteorologist on a voyage to Portugal and Madeira.[10]

Tor Bergeron used the metaphor of ski jumping to describe Rossby's character. Once on a skiing trip, a youthful Rossby (who was by no means athletic and not a trained skier) persevered until he mastered a K-47 meter ski jump at Lilla Fiskartorpet. According to Bergeron, many, if not most, of Rossby's decisive steps in life as a man and a scientist "were daring jumps into the unknown that succeeded, and admirably so."[11] Such was his leap to Bergen. To invoke another metaphor, the influence of Bjerknes diverted Rossby into a career in meteorology but did not capture his sole allegiance. Rossby was much too independent for that. He was more like a brilliant comet on a wildly eccentric path into the unknown than an inner planet occupying a stable orbit in the inner circle of the Bergen school.

Rossby's First Encounter with the U.S. Weather Bureau

Rossby won an American-Scandinavian Foundation fellowship for 1926 that brought him to the U.S. Weather Bureau "to study dynamic meteorology problems." One of Rossby's goals included demonstrating the applicability of the polar-front theory to American weather. Anne Louise Beck in 1922 and Jacob Bjerknes in 1924 had attempted similar incursions.[12] Yet Rossby was not simply a disciple of the Bergen school. Like Vilhelm Bjerknes, he was a charismatic leader, a dynamicist with big theoretical ideas, an intuitive and innovative physical thinker, and a builder of institutions. During a very active first year in America, he studied convection and turbulence, consulted on aviation safety, and reexamined U.S. weather maps based on the theory of the polar front.

Rossby's first project was to construct a rotating-tank experiment in the basement of the Weather Bureau, scaled to emulate the atmosphere. In this, he was inspired by Helmholtz's hydrodynamical theory of vortex motions. In 1857, Helmholtz, possibly at his own dinner table, described fluid vortices generated by drawing a spoon rapidly for a short distance along the surface of a liquid—perhaps in his teacup or his soup bowl—and then quickly drawing it out: "The vortex rings advance, broaden when they encounter a [boundary], and are enlarged or diminished by the action of other vortex rings precisely as we have deduced from the theory."[13] Many others, including Vilhelm Bjerknes in his youth, had proposed or built qualitative hydrodynamic models, and Felix Exner had recently described and illustrated a rotating-dishpan experiment in the second edition of his *Dynamische Meteorologie* (figure 3.1). Rossby's tank was two meters in diameter and was filled with colored fluids of different densities; it rotated around a vertical axis three to four times per minute. Like the atmosphere, the tank had large horizontal dimension compared to its vertical depth. To a first approximation, it was two-dimensional. This was the key to Rossby's theoretical approach. Although the tank suffered mechanical failure and his initial attempt to emulate the atmosphere and write nondimensional equations of its motion was inconclusive, some two decades later this line of research eventually proved fruitful.[14]

Rossby also collaborated with Weather Bureau forecaster Richard H. Weightman on two articles on polar-front theory in the United States. Junior meteorologist Hurd C. Willett helped with the maps. Much like

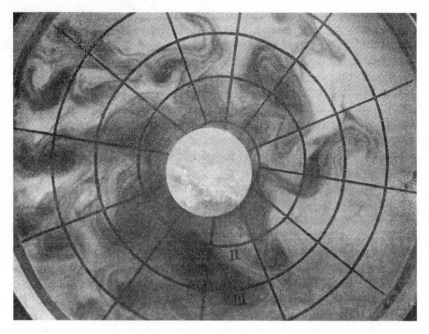

Figure 3.1
Felix Exner's dishpan experiment. Photograph of a rotating tub filled with water and stained with dye with a cylindrical piece of ice in the center and heated at the edges. The waves formed in the tank emulate the circumpolar circulation of the atmosphere. *Source:* Exner, *Dynamische Meteorologie*, 2nd ed., 341.

Anne Louise Beck and Jacob Bjerknes had done several years earlier, these papers analyze the development of a complex weather situation over several days, concluding that a denser network of surface and upper-air stations with better and more frequent observations was needed for reliable analysis of both the positions and the energies associated with the storms. This was especially true over the vast western mountain ranges, where fronts often disappeared only to reform downstream, and in the moist and conditionally unstable air masses of the Gulf Coast, where conditions were significantly different than in Norway. Even without the needed improvements in the observations, Rossby and Weightman concluded, optimistically, that their study had provided "conclusive evidence" that "the polar front theory can be applied with great advantage to even rather complicated weather maps in the United States and that it enables us to explain phenomena which without a knowledge of the dynamics of the situation would hardly be understood."[15]

The Model Airway, Marvin's Weather Bureau, and MIT

Aviation got a boost in 1926 when the Daniel Guggenheim Fund for the Promotion of Aeronautics dedicated $2.5 million "to promote aeronautical education throughout the country; to assist in the extension of aeronautical science; and to further the development of commercial aircraft, particularly in its use as a regular means of transportation of both goods and people." The goal was to assist in making air transportation "safe, popular, and regularly available."[16] Wilbur Wright was known for his speeches emphasizing lift, power, and control as the three general classes of obstacles to be overcome in building a successful flying machine. Weather was the fourth obstacle to be faced while operating it. Bjerknes knew this and had demonstrated the physical equivalence of the forces that lift airplanes and the forces that shape the weather. Rossby, working with the Guggenheim Fund, aimed to apply these lessons to the burgeoning aviation industry and make the skies safer for flying.

The timing was propitious. The Air Commerce Act of 1926 established a bureau of aeronautics to regulate the operation of airports and to license aircraft, engines, pilots, and related personnel. It also provided support for aviation by establishing regulations for inspecting aircraft and providing communication facilities and navigational aids for all flying activities on federally established civil airways. The act directed the Weather Bureau to "furnish weather reports, forecasts, warnings, to promote the safety and efficiency of air navigation in the United States" and reassured insurance companies, private investors, and banks that safety standards would be enforced. The National Advisory Committee for Aeronautics called it "the legislative cornerstone of commercial aviation in America."[17]

Rossby's ideas on improving the meteorological side of aviation made a big impression on Harry Guggenheim. When Rossby's American-Scandinavian fellowship ended, Guggenheim provided him with funding, and the growing needs of aviation provided him with both duties and opportunities. The two men had been introduced by Lieutenant Francis W. Reichelderfer, the thirty-one-year-old chief of navy aerology.[18] "Reich" studied Bergen school methods on his own and enjoyed conversations with Rossby in the Weather Bureau map room, where he frequently prepared charts for the navy.[19]

In 1927, Rossby was appointed chair of the Guggenheim Interdepartmental Committee on Aeronautical Meteorology, whose members included Willis R. Gregg from the Weather Bureau, William R. Blair from the U.S. Army, Thomas F. Chapman from the Department of Commerce, and Reichelderfer from the U.S. Navy. All were senior to Rossby, some by multiple decades. The committee was tasked with promoting mutual understanding between pilots and meteorologists and supporting the scientific study of the weather. To promote aviation, Rossby helped with forecasting for the transatlantic flight of Commander Richard E. Byrd and his crew. The fund also supported the thirty-five-city tour of Floyd Bennet, Byrd's pilot during the North Pole trip, and the eighty-two-city tour of Colonel Charles Lindbergh, who was a trustee of the fund. He also supported Weather Bureau colleague Hurd Willett's fellowship to study Bergen methods for a year in Norway.

Legend has it that a "series of minor incidents" in Marvin's Weather Bureau made Rossby "persona non grata" there. This is only approximately true. One embarrassing incident surrounded Lindbergh's planned capital-to-capital flight from Washington, DC to Mexico City in the *Spirit of St. Louis*. Lindbergh had asked Rossby for "a full report on flying weather in the southern states" for a planned takeoff on December 14, 1927. Rossby forecast unsettled conditions all week, "with head winds which would make a nonstop flight impossible." Lindberg then announced in the press a ten-day delay—it seems, to keep the crowds away from Bolling Field—and then, ignoring the forecast, he took off "by-the-seat-of-his-pants" early on December 13 on his twenty-six-hour flight. So much for Rossby's forecast. Chief Marvin was incensed when the newspaper billed Rossby as "chief meteorologist of the Washington Weather Bureau" and retaliated by banishing him from the bureau's 24th and M Street headquarters.[20]

By then, Rossby had other options and focused his attention on the Guggenheim project to initiate a model airway weather reporting service in California. Similar to Bjerknes's recruiting in 1917–18 of farmers, fishers, and lighthouse keepers for the West Norway Weather Service, in early 1928 Rossby traveled along the flyways between San Francisco and Los Angeles recruiting observers for the model service. By May, the system was in place

Figure 3.2
Horace Byers, a college junior in 1928, at the blackboard at the Oakland Airport,
issuing forecasts and recording upper-air wind measurements for the model airway.
Source: National Archives and Records Administration, RG 27-G-2-4-25545-C.

with cooperation from Western Air Express, Pacific Telephone and Telegraph, the air services of the army and navy, the Department of Commerce,
and the U.S. Weather Bureau. Edward H. Bowie, meteorologist in charge of
the Weather Bureau's San Francisco office, assumed responsibility for the
system in July and supervised the training of observers and forecasters, one
of whom was Horace Byers (figure 3.2), a junior at the University of California, Berkeley.

As Bowie explained it, the concept was simple: just as good paved roads
were necessary for the automobile to become a commercial success, so too
were "aerial roads"—that is, accurate weather forecasts for safe flying
conditions—needed for the growth of commercial and military aviation.
For several decades, roads for automobile use had expanded to reach into
remote areas with generous funding from municipalities, counties, states,
and the nation. So, too, Bowie argued, public funds should be provided to
enhance public confidence in aviation and routinize air transport in a

medium (the atmosphere) that can be at times "as quiet and placid as a mill pond and at other times as agitated, turbulent and treacherous as the rapids below Niagara Falls"; at times as transparent as crystal and at other times opaque, "like pea soup." The model service set out "to demonstrate the gains in flying hours and safety to be derived from a proper utilization of complete meteorological information." Bowie claimed there were collateral benefits. Telephone weather reports provided to the general public could improve highway safety, promote tourism, assist growers, and help with responses to forest fires.[21]

The Guggenheim model airway, later expanded across the nation by the federal government, aimed to routinize air travel and discipline pilots. California was chosen because of its ever-increasing amount of commercial, military, and private air traffic, the lack of competing services, the cooperation offered by the state, and the strategic interests of the Fund, which was investing in aeronautical engineering at Caltech. The topography is complex, storms arrive off the Pacific Ocean unannounced, and fog frequently blankets the bays, valleys, and nearby mountains: all were ample challenges for the meteorologists involved. Air mail pilots felt comfortable with about 300 feet of clearance between the mountains and the cloud base but, failing that, flew blind into the clouds—climbing, estimating the terrain, counting the seconds, and hoping that a subsequent dive would take pilot and plane safely into the next valley, not into the mountainside. At the time, most flying was seat-of-the-pants, with many pilots claiming that their personal experience in an open cockpit exposed to wind and rain was the best and most reliable weather instrument.

Weather Bureau chief Marvin emphasized service, discipline, and the routine task of issuing the daily forecast, but he was also an active researcher who specialized in the design, construction, standardization, and use of meteorological instruments, particularly large box kites and aerometeorographs. He had theoretical interests regarding the phenomenon of inertia motion and wrote about the "law of the geoidal slope" in relation to the Coriolis force. Marvin was a founding member of the National Advisory Committee for Aeronautics, member ex officio of the International Meteorological Organization, and president, in 1926, of the American Meteorological Society. The Great Depression was not kind to Marvin's Weather Bureau. Reductions in public service in 1933 included cuts in staffing and pay; a reduction in the number of flood and storm warnings, crop reports, daily telegraphic forecasts, and weather maps; the demotion of two dozen first-order stations to cooperative status; and the closing of many others, including some in the fledgling airways weather service.[22]

Marvin had met Vilhelm Bjerknes several times and remained informed about his work but was suspicious of the exchange of personnel and resistant to meteorological proselytizing. Nevertheless, Marvin had offered Anne Louise Beck a position in 1922 and welcomed Jacob Bjerknes's visit to the bureau in 1924. Marvin wrote a strong letter of recommendation for Rossby in 1927, and sent it, along with Rossby's biographical sketch, list of publications, and several reprints, to MIT's president Samuel Wesley Stratton. Marvin referred to Rossby's "exceptional training in meteorology, both theoretical and applied," the effort the Weather Bureau had made to extend his fellowship, and the grant they had secured for him from the Guggenheim Fund to retain his services, but said that they could not offer him a permanent position because he was a Swedish citizen. Marvin thought MIT might profit from the services of this young man. The Weather Bureau had long hoped to see meteorology taught in American universities as a branch of physics and mathematics rather than geology and geography. Marvin said "he knew of no young man, whether in this country of abroad, better equipped to teach meteorology as it should be taught." Weather Bureau physicist William J. Humphries sent a similar letter of recommendation.[23]

Stratton, who served as the first director of the National Bureau of Standards and the president of MIT since 1923, was interested and wrote back, asking about Rossby's salary and rank expectations. Marvin replied that Rossby's contract with the Guggenheim Fund for the Promotion of Aeronautics had just been extended for six months at $250 per month. Soon after, Stratton met Rossby in Washington, DC at a meeting of National Academy of Sciences where he "formed a very favorable impression of him" and was considering the possibility of employing Rossby as an instructor in meteorology or perhaps hydraulics at MIT.[24] Trying to add a little urgency and with the Guggenheim appointment in hand, Rossby informed Stratton in early July that he had received a "tentative offer" of employment at another institution and would appreciate an early reply from Stratton. Stratton's letter disappointed Rossby:

I regret that there is no vacancy here in which you would be interested. I had hoped that we might introduce some work along the lines of meteorology in our Department of Physics, but other urgent needs for the expansion of the work in that Department make it seem inopportune to take up any new work at the present time. I am very sorry indeed that the Institute is not in a position to avail itself of your services at this time.

In Stratton's opinion, an instructor in meteorology seemed "a little bit like a luxury rather than a necessity."[25]

The situation was much different a year later based on a request from the naval postgraduate school for a course in aerology at MIT. The presence of six navy students tipped the scales, prompting Stratton to authorize a new course in meteorology and seek funding from the army, the navy, and Guggenheim to hire Rossby.[26] Stratton asked the Daniel Guggenheim Fund for the Promotion of Aeronautics for assistance in establishing a meteorological course within the aeronautics department "for the advanced training of service and civilian students at the Institute, with special reference to aeronautical meteorology." He asked for compensation for Rossby, who would be appointed with faculty rank, $10,000 for equipping and maintaining a laboratory, and two research fellowships per year. The wheels were turning. Harry Guggenheim acknowledged the letter almost immediately and promised to bring it to the attention of the members of the fund at their next meeting. Edward P. Warner, assistant secretary of the navy for aeronautics, spoke to Guggenheim and suggested that Rossby be made an associate professor.[27] A letter dated June 18, 1928 delivered the good news—$34,000 from the Foundation to fund the new program for three years, which included $10,000 for equipment, $12,000 for Rossby's salary, $6,000 for three fellowships, and $6,000 for research.[28] Rossby reported for duty at MIT on September 7.

Rossby's MIT Program

It should always be remembered that the weather map is the fundamental meteorological laboratory, even more essential to meteorological studies than the wind tunnel is to an aeronautical laboratory.
—Carl-Gustaf Rossby, 1928[29]

Rossby argued for a "modernization of American meteorology" starting in New England. He proposed a dense network of twenty stations connected by telephone or telegraph covering the northeastern states and taking observations four times per day. With this move, he attempted to emulate the *Vervarslinga på Vestlandet*. He also wanted upper-air soundings at Boston

to altitudes of eight to ten thousand feet. Students should collect and analyze their own current data: "Without such organization, all synoptic work
is bound to become guesswork." Pointing to his 1926 article with Weightman and the need to apply polar-front theory to American weather maps,
he proposed bringing one of Bjerknes's young protégés to MIT for a period
of six months. The cost of this program, over and above the Guggenheim
funds, was $39,600 the first year and $16,600 per year after that.[30]

It was a busy first year. Rossby got some of his budget approved but not
the twenty New England stations. Willett, fresh from his fellowship in Norway, joined the MIT faculty as an assistant professor. He taught synoptic
meteorology and the meteorological laboratory course (weather maps,
cyclones, cyclone family, polar front, general circulation, air mass, and
frontal analysis) using articles by Jacob Bjerknes, Solberg, and Bergeron and
wrote his own hundred-page treatise on dynamic meteorology for a volume
published by the National Research Council.[31] Rossby taught dynamic
meteorology and hydrodynamics using the works of Vilhelm Bjerknes,
Felix Exner, and Sir Horace Lamb. He also ran the meteorology seminar
where recent papers were discussed, supervised student thesis research and
writing, and began the "Professional Notes" series of in-house publications.
He received funding for aircraft soundings and instrument calibration and
played host to his colleague "Reich" who visited MIT in December "to get
the airplane soundings down at the Boston Airport started, and get that station running regularly."[32] Sensitive to the needs of aviation as expressed by
Guggenheim, Rossby proposed an experimental program on fog at the new
Round Hill research facility overlooking Buzzard's Bay.[33] The study and clas-
sification of American air masses began in 1929, with special attention to
their source regions, characteristics, and vertical structures. At the start of
his second academic year at MIT, Rossby married Harriet Marshall Alexander, the daughter of a Boston physician. The couple had three children.

Rossby moved fast and aggressively sought to establish new programs
and request new funding from MIT sources. He was well connected and
served on a number of prominent national aeronautical committees, which
provided him with plenty of new ideas.[34] President Stratton's handwritten
note to the head of the aeronautics program indicates the pace and the
strain of trying to keep up with Rossby: "Tell Rossby that the Guggenheim
and Green items are OK. The other [MIT] items have been approved by me,
will go to the Ex Com next month, and will no doubt be approved. He can
go ahead on that, but *never* send out an official budget or personal notice

until approved."[35] Rossby submitted a flurry of proposals for more cooperative work—at Round Hill, with Harvard and Blue Hill, on air pollution, and
with oceanographers. Impatient, petulant, and seeking an expression of
trust from Stratton for one of his proposals, he even threatened to resign.
Stratton had to remind him that he was getting almost everything he had
requested but had to work with the system: "Do not make any proposal
until you have taken it up with me." He advised Rossby to slow down the
pace: "We are thoroughly interested in the meteorological work and expect
to build it up. We may not be able to do so quite as rapidly [as you would
desire]."[36]

Rossby had a solid supporter and a fan of aviation in the president's
office. When the new Goodyear-Zeppelin Corporation blimp *Mayflower*
arrived at the Institute drill grounds in July 1929, Stratton took an hour's
ride over Boston and vicinity. He recommended that Rossby "fit it up for
meteorological studies, perhaps with radio for the fog work."[37] He also
funded Rossby's 1930 summer trip to Europe.[38] The itinerary—to Frankfurt
am Main, Vienna, Darmstadt, Leipzig, Berlin, Hamburg, Bergen, and
Stockholm—included research and consultation on dynamic, synoptic, and
maritime meteorology, radiation phenomena and turbulence, forecasting
and instructional practices, upper-air observations, and attendance at the
International Geophysical Union meeting. The scope and intensity of the
trip reflected Rossby's desire to deepen his theoretical understanding of
hydrodynamics, catch up with the latest forecasting practices, and perhaps
most important, begin to heal the rift he had experienced firsthand between
the Austro-German and Norwegian schools. His goal was to resolve the
intellectual differences between Exner and the Bergen school regarding
links between the stratosphere and weather nearer to the surface. In his
1930 annual report, Rossby firmly linked the course in meteorology at MIT
to the safety of aerial navigation. This required revolutionizing forecasting
practices by emphasizing the relevance of theoretical meteorology. He
argued that MIT was the only institution in the United States where work
of this type was being carried on successfully. By age thirty-one and after
only five years in the United States, Rossby had become an established figure in American and European meteorological circles—but not yet because
of his theoretical contributions.[39]

In 1930, the distinguished physicist Karl Taylor Compton assumed the
presidency of MIT. He continued to nurture the meteorology program,

adding to the staff K. O. Lange, an instrument specialist from Darmstadt, and Daniel C. Sayre, a pilot from the aeronautics course.[40] The research seminar was expanded to include a direct collaboration with the new Woods Hole Oceanographic Institution and publication of a new series of "Papers on Physical Oceanography and Meteorology," with early contributions on "American Air Mass Properties" by Willett, "Dynamics of Steady Ocean Currents in the Light of Experimental Fluid Mechanics" by Rossby, and "A Study of the Circulation of the Western North Atlantic" by oceanographer Columbus Iselin. The joint MIT-Woods Hole program on the dynamics of oceans and atmospheres, established by Rossby and the biological oceanographer Henry Bryant Bigelow, emulated the cooperation three decades earlier in Norway among Fridtjof Nansen, Bjørn Helland-Hansen, Vilhelm Bjerknes, and their associates.[41]

With grants from the National Academy of Sciences, Guggenheim, and MIT, the meteorological program purchased a research airplane in 1931 to carry aloft a meteorograph (figure 3.3), a device that made a continuous record of temperature, relative humidity, and barometric pressure.[42] Sayre piloted the MIT plane to altitudes exceeding 5 kilometers over Boston, even under adverse weather conditions. Similar to the pioneering but ill-fated program of Ernst Calwagen in Norway in the early 1920s, the aerological information was used in connection with synoptic studies and local investigations of the turbulent layer near the ground. The data collected also provided students with opportunities to make real aviation forecasts.

Rossby focused his program at MIT on the study of thermodynamics applied to air mass analysis. He was able to link the upper-air soundings taken by airplane to the identification of air masses and their changes over time. By plotting the data on a thermodynamic diagram, he could identify areas of instability susceptible to the formation of fog, clouds, and precipitation. He and his students developed the principles of isentropic analysis, replacing vertical pressure coordinates with surfaces of equal potential temperature on which to trace dynamical properties. Using these techniques and information from newly available experimental radiosonde ascents, the MIT scientists were able to expand their analysis beyond particular weather systems and apply thermodynamics and fluid mechanical principles to the general circulation of the entire atmosphere. Rossby also studied friction and turbulence in the boundary layer, both over the ground and over the sea surface, and succeeded in adapting concepts from

Figure 3.3

The aerometeorograph, a self-recording instrument used on aircraft to chart simulta-neous values of temperature, atmospheric pressure, and humidity, was quite similar in design to this Marvin kite meteorograph, ca. 1913. *Source:* National Archives and Records Administration RG-27, 2 27-G-5–5-10744-C.

aerodynamics to meteorology and oceanography, providing a corrective to the idealized circulation theorem of Bjerknes. The pace of his ideas was relentless. He moved from studies of small-scale interactions near the sur-face to large-scale mixing in the atmosphere, to frictionless flows in the upper wind currents, to interactions between the stratosphere and the tro-posphere. The MIT meteorological program hosted Jacob Bjerknes, who gave a series of twenty lectures in 1933 on the general circulation of the atmosphere and convened an informal conference on fundamental meteo-rological problems with representatives from the U.S. Navy, U.S. War Department, and U.S. Weather Bureau.

German-born meteorologist Bernhard Haurwitz arrived at MIT in 1932 on a short-term research fellowship. He was trained in meteorology by Ludwig Weickmann in Leipzig and in geophysical hydrodynamics and forecasting by Vilhelm Bjerknes and his associates in Oslo and Bergen. Bjerknes introduced him to Rossby, and Rossby found Haurwitz a fellowship, connected him with the Blue Hill Observatory, and at times, supported him out of his own pocket. The political situation in Germany was such that Haurwitz never returned; he worked in the United States and Canada for the rest of his life. He liked the informality of MIT and the "very lively" graduate students he instructed, including Wexler (the subject of chapter 4). He recalled fondly, "When summer came, we all went down to Woods Hole and stayed there and worked with the oceanographers. Some of us did meteorology, so it didn't really make too much difference at that time if you were a meteorologist or oceanographer." [43] When Haurwitz lectured at the Carnegie Department of Terrestrial Magnetism in Washington on the perturbation techniques he had developed with Bjerknes and Solberg, he was introduced as "the only man in this country familiar with and actively participating in this phase of the Norwegian work."[44]

By 1935, Rossby was "well on his way" to establishing a new school of meteorology. His was "a second great meteorological advance," grounded in the theoretical foundations of Vilhelm Bjerknes and the practical techniques of Bergen but transcending them and going beyond the polar front concept to emphasize hemispheric flows, planetary waves, and ultimately jet stream meteorology. Wherever he went, he was the center of a moving seminar. He was approachable, charming, and open to new ideas and theoretical discussions. Bergeron recalled his "innumerable followers and friends," while Byers commented on his exhilarating and inspirational manner.[45]

"A Time of Transition"

It is always "a time of transition" somewhere, but in meteorology, some of the changes at the U.S. Weather Bureau in the mid-1930s were exceedingly important and noteworthy. In his biographical sketch of Rossby for the memorial volume, Tor Bergeron mentioned in passing that "air-mass methods were officially introduced into American weather service in 1930."[46] Almost, but not quite. That occurred mainly after 1934 under the leadership of Willis R. Gregg, and the process was completed after 1938 by chief Francis W. Reichelderfer, who had spent time studying in Bergen. The official Weather Bureau pamphlet, "Weather Forecasting from Synoptic Charts," written by Beck's old nemesis, A. J. Henry, paints a dismal picture of the situation in 1930. The Weather Bureau at the time relied not on air mass analysis but on pressure tendencies and "the experience and natural ability of the forecaster." Henry provided a brief and superficial description of Bergen methods but couldn't resist a swipe: "In many respects the new conception harks back to Dove, who in the middle of the nineteenth century pictured the weather as being due to the conflict between currents of equatorial and polar air, respectively."[47] The next edition of this pamphlet, issued in 1936, tells a much different story.

By 1933, based on the daily soundings and reanalysis of national weather observations, MIT was mailing *to* the U.S. Weather Bureau "the only published daily analyses of American air masses." Rossby had insider information that the meteorological subcommittee of the Science Advisory Board was planning to recommend the adoption of air mass analysis. He wrote in his annual report that he had "reason to hope" for a "general reorganization" of the Weather Bureau within the year along with the need to hire a large number of MIT meteorologists trained in this technique.[48] The Science Advisory Board had been created by presidential order in 1933 as the scrutineers of science in the federal government. Robert A. Millikan of Caltech chaired the special committee on the Weather Bureau, whose members included National Research Council chair Isaiah Bowman, MIT president Karl Taylor Compton (Rossby's boss), and Charles D. Reed, a senior meteorologist from Iowa. Their report, issued late in 1933, contained two major recommendations: (1) extend the air-mass analysis method of forecasting, and (2) consolidate the whole national system of recording and reporting meteorological data under the Weather Bureau. They also recommended improving the quality of observations, the frequency of forecasts, and the training of Weather Bureau personnel "in the more modern methods." The committee advised that the transition to a new system of frontal and air mass analysis be made with caution to avoid jeopardizing the present valuable services.[49] Based on the recommendations of the board, Willis R. Gregg took the oath of office as chief of the U.S. Weather Bureau on January 25, 1934. His administration, if it is studied at all, is often linked to his predecessor, Charles F. Marvin, who has been caricatured as a recalcitrant bureaucrat who stifled the adoption of modern forecasting methods and had to be removed from office by the National Science Board.[50] The historical record, thin as it is, reveals that this is only partially true. Marvin was not forced out of office; he was seventy-five years old when he retired as Weather Bureau chief in 1933.

Tradition has it that nothing much happened until 1938, when Reichelderfer became chief, but evidence shows otherwise. Gregg was a recognized leader in the new field of aerology, he served with Rossby on the Guggenheim committee on aerological meteorology, and, in the depths of the Great Depression, he began the process of adopting forecasting methods using air mass, frontal, and isentropic analysis. In 1922, under the auspices of the National Advisory Committee for Aeronautics, Gregg published the first "standard atmosphere"—a hypothetical vertical distribution of atmospheric temperature, pressure, and density that, by international agreement, is taken to be representative of the atmosphere for applied aviation purposes. He was confirmed as chief of the Weather Bureau during a period of federal budget cutbacks, Science Advisory Board oversight, and rapid conceptual and technological change in meteorology.[51] Gregg accepted the board's findings and agreed that, due to the lack of trained personnel, budget constraints, a dearth of observations, and institutional inertia, the transition to air mass analysis at the Weather Bureau might require three to five years to implement fully. Nevertheless, he began immediately. His appointment book for his first month on the job, records consultations with Hurd C. Willett, Jacob Bjerknes, the Science Advisory Board, and Rossby.[52] He announced he was adding to the staff "several well-qualified men who have specialized in forecasting based on air-mass analysis." He also initiated an airplane observational program with the interdepartmental cooperation of MIT and the Departments of the Army, the Navy, Agriculture, and Commerce. By July 1934, twenty stations were collecting data necessary for the new forecasting techniques, and Gregg announced the start of a new "three dimensional meteorology" of air mass analysis. Rossby responded gleefully, "With the official acceptance by the United States Weather Bureau of the method of air mass analysis, the first objective of our work has been attained"—but only the first. [53] The next step was hiring and training new personnel. Horace Byers received his ScD in meteorology from MIT under Rossby in 1935 and joined the Weather Bureau as an associate meteorologist in charge of the new air-mass analysis section.[54] Working with Byers were Harry Wexler, Stephen Lichtblau, and junior forecaster Charles H. Pierce. Their average age was twenty-five. This new "theoretical" group initially encountered rather stiff resistance, They held public map discussions late in the morning after the "practical" forecast had been issued by traditional methods. The young "upstarts" gained plenty of experience by defending their "new-fangled" ideas before senior forecasters twice their age.[55]

The Weather Bureau in-service training program sent many of its forecasters back to school, some of them to earn advanced degrees, using funds provided by the Bankhead-Jones Act of 1935 for agricultural research. In 1935–36, Norwegian meteorologist Sverre Petterssen, a Bjerknes acolyte and head of the Bergen weather center, visited America to lecture for the navy and Weather Bureau on weather analysis and forecasting. Wexler attended and kept an extensive set of notes. These lectures became the core of Petterssen's books *Weather Analysis and Forecasting* (1940) and *Introduction to Meteorology* (1941). Gregg admired Petterssen and offered him employment as the deputy director in charge of a program of research, education, and scientific services, a position later held by Rossby. Petterssen declined the offer but thought that Gregg was "an able administrator with progressive outlooks."[56]

The 1936 edition of "Forecasting from Synoptic Weather Charts," when compared with its 1930 counterpart, reveals the sea change that had occurred during Gregg's tenure as head of the Weather Bureau. Rossby's colleague Richard H. Weightman, the new author, began, "Weather forecasting, born in empiricism, is slowly but surely developing along sound physical lines. The introduction of frontal methods, the later detailed analysis of the air masses, and the more recent development of kinematical methods for forecasting the movements and intensities of fronts, highs, lows, etc., are meteorological milestones in the path of progress."[57] Although Henry's 1930 text had mentioned the Bergen school only in passing and even then quite dismissively, the majority of Weightman's 1936 report was based on Bergen methods and contained twenty illustrations of Norwegian-style analysis.

Radiosondes

Communications are the Alpha and Omega of Meteorology.
—Carl-Gustaf Rossby, 1929[58]

"The possibilities for the use of the principle of the radio-meteorograph or the 'robot observer' are, in fact, almost endless and abundantly justify all the feverish activity that is now in evidence in many countries in bringing the equipment to a point where it can be used in regular service." So wrote Willis R. Gregg in 1936.[59] The new technology of the radiosonde was catching on: a free-flight balloon carried an expendable meteorological instrument package aloft to measure, from the surface to the stratosphere, the vertical profiles of temperature, pressure, wind, and humidity and transmit the data via radio to a receiver-recorder on the ground.

Previously, direct methods of probing the atmosphere involved sending devices called "meteorographs" (figure 3.3) aloft on kites or balloons to measure and record variables such as temperature, pressure, and humidity on a rolling paper chart. Early balloonists took maximum and minimum thermometers with them on their ascents; others attached instruments to kites. In 1892, in Paris, Gustave Hermite and Georges Besançon inaugurated a widespread practice of using balloons to lift meteorographs; a decade later Léon Teisserenc de Bort in France and Richard Assmann in Germany announced that their instrumented balloons had measured two new layers of the atmosphere—the tropopause, where the temperate was constant, and the stratosphere, where the temperature increased with height. Regular observations using instrumented kites, tethered balloons, and visually-tracked pilot balloons provided most of the aerological data in the first two decades of the twentieth century. For a brief period in the late 1920s and early1930s, aerometeorographs were carried aloft by aircraft.[60] With the exception of a few experiments using kite wire to transmit data, these methods inevitably involved delays between collecting and interpreting the measurements, and in some cases the balloon-borne meteorographs were lost completely.

The radiosonde is "one of the more significant technological innovations of the twentieth century," wrote John DuBois and his colleagues in *The Invention and Development of the Radiosonde*, "not only because its widespread use greatly enhanced the accuracy of weather forecasting, but also because some features of its basic design became the foundation of all modern analog telemetry systems." Basic research on balloon-borne radio transmitters began in 1921 at the Lindenberg Observatory under the supervision of Hugo Hergesell, and a major breakthrough occurred in the United States when in 1924, Colonel William R. Blair, who had lost sight of many pilot balloons in the war during their ascent into clouds, attached a temperature sensitive buzzer to two tandem balloons and tracked them successfully by radio signals for about twenty minutes, retrieving both wind and temperature data. According to DuBois, "Blair's device constituted a radio-sonde, although this terminology was not introduced until 1931."[61] Another thread dates to 1927 in France, where Robert Bureau and Pierre Idrac invented the radio frequency (RF) tube-type transmitter and, in 1929, flew their radiosonde, called the "Thermoradio," equipped with a bimetal-lic temperature sensor. DuBois and his colleagues called this instrument "the first true radiosonde." Two other models quickly followed—the Moltchanov radiosonde in Russia and the Duckert radiosonde in Germany, the latter being the first to measure pressure, temperature, and humidity. The next steps involved making the devices more reliable and cost-effective for the use of researchers and weather services. Finnish entrepreneur Vilho Väisälä established a company to produce radiosondes in 1936 and deliv-ered the first commercial order of twenty of them to Carl-Gustaf Rossby for research use at MIT.[62]

Gregg witnessed his first radiosonde launch on September 20, 1935, at the Slutsk Observatory near Leningrad while on an official visit to the Soviet Union. After a light luncheon and a vodka toast, the afternoon included "a special balloon flight with a radio-meteorograph attached." He was able to "listen to the signals as they came back from the balloon and watch the computation and charting of the data."[63] Two years later, the U.S. Weather Bureau launched its first radiosonde at East Boston, Massachusetts, and soon brought them into regular practice. Piloted aircraft soundings were phased out because balloons could fly much higher than aircraft and could be launched in virtually all weather conditions with no risks to avia-tors. A radio receiving ground station was part of every radiosonde launch (figure 3.4).

A wave of technological enthusiasm crested in 1938. As Gregg was pre-paring his presidential address for the American Meteorological Society, he asked his colleagues to imagine what the meteorological profession might look like in fifty years. Visions of the future focused on existing communi-cations technologies such as radio and the radiosonde and the possibilities of television, "robot reporters," rockets, satellites, and even weather con-trol. Reichelderfer, one of the respondents, thought it was presently both desirable and possible to cover the wide-open ocean spaces by use of daily reports from a radio robot, either tethered or slowly drifting with the cur-rents. The upper-air robot (radio meteorograph) might soon replace the present surface weather map, and fronts would be announced immediately

Figure 3.4

A radiosonde ready to be launched. The instrument package with parachute is attached to a sounding balloon, and a ground station receives the radioed data. *Source:* National Archives and Records Administration RG-27, MD 27-G-10-16-1.

as they pass a station, not "diagnosed" later by indirect map analysis. George W. Mindling foretold, in doggerel, of the "coming perpetual vision-tone show" of perfect surveillance and perfect prediction using television and infrared sensors.[64] Meteorology was becoming more electronic. With the addition of a radio direction finder, radiosondes became rawinsondes, equipped with transmitters that made it possible to measure high-altitude wind directions and speeds. The meteorological radiosondes of the 1930s provided significant technological models for more specialized remote sensing applications such as cosmic ray, ozone, and ultraviolet radiation detectors, and eventually, radio telemetry with rockets and satellites. Training in electronics was becoming an essential part of meteorological education. In 1938, in his fourth year as chief, Gregg died of a heart attack at age fifty-eight. It was left to his successor Reichelderfer to complete the reforms. Rossby and his associates had brought air mass analysis to the U.S. Weather Bureau, but there was much more to do.

Rossby in Charge of Research at the Weather Bureau

When Francis W. Reichelderfer became Weather Bureau chief on December 15, 1938, he placed training programs involving frontal and air mass analysis center stage. His predecessor Gregg had begun the process by establishing the Air Mass Analysis Section, but Reichelderfer still faced stiff internal opposition from the traditionalists during the early years of his tenure. Rossby wrote a long letter to Karl Taylor Compton about the situation:

At present there is a terrific struggle going on in the Bureau for power. Too many of the well-meaning old hands are in power, and we are definitely missing the opportunity to modernize the Bureau quickly and effectively. In my opinion Cmdr. Reichelderfer is a bit scared of the new crowd and the result is that men like (former acting director) Charles C. Clark and the others are very much in power.[65]

To promote reform, Rossby agreed to serve for three years as assistant chief of the Weather Bureau in charge of research and education, arriving in Washington on July 1, 1939. Petterssen replaced him at MIT as chair of the meteorology program. Rossby immediately regretted the move. Within three months he was expressing his dissatisfaction to Compton: "Personally, I wish to resign from my position in the Bureau on Dec. 31, 1939 or Jan. 31, 1940 and obtain a sabbatical leave for a full year or else a research professorship at MIT."[66] He found himself flush with new ideas, but was stifled by the massive bureaucracy and burdened by the training program initiated by Reichelderfer, which included cycling up to 180 Weather Bureau staff members each year through three-month crash courses in Washington, DC, Chicago, and Oakland. Ten Weather Bureau employees per year were also to be sent to universities. This emphasis on service and teaching was coming at the expense of research, and Rossby thought his presence in Washington, DC, was not needed; he could supervise the education program from Boston.

Rossby was aware that the Bergen school was preparing, "with ample support from the Carnegie Foundation," volume 3 of the long anticipated *Dynamics*, and he too wanted to prepare a readable and coherent book containing the "very fine results" of his own group and a formulation of the fundamental problems yet to be solved:

I dare say that we have created many more new and fertile ideas than the Norwegian group, and under much more strain. I am desperately anxious to have these ideas organized into a readable, coherent discussion of the atmosphere, a discussion summing up the work to date and formulating the fundamental problems to be solved. This could be done in a year, provided I had one good assistant at my disposal and no other duties. . . . I am the only one who can write that book, but there must be others who could be put at Reichelderfer's side to help him and who would do a better job than I can do.[67]

Compton advised Rossby that, under these circumstances, it would be "wise and proper" for him to resign because "the opportunity to create a great research center in the U.S. Weather Bureau apparently does not exist in the near future." He believed that the Weather Bureau's policy of focusing only on forecasting and other services would be "fatal" if continued too long. It would eventually result in obsolescence of the service and demoralization of the personnel. He suggested to Rossby that perhaps a vigorous research program could be undertaken, "just as soon as the routine procedures of the new methods can be got into smooth running operation." He thought Rossby should exit the Weather Bureau on a positive note, promise them his future assistance, and return to MIT to continue his research and writing. He was trying to find grant money for Rossby to do this, but it was clear that MIT could not offer him a salary if he returned midyear.[68] Rossby didn't leave the Weather Bureau and never wrote a book. He decided that the field was changing too quickly for that and that any book would soon be out of date. Instead, he produced several landmark articles that summarized the progress to date.

The MIT School of Meteorology

In 1941, Rossby published a fifty-page article called "The Scientific Basis of Modern Meteorology."[69] It was an accessible summary that covered a vast array of issues, mostly centering on understanding the general circulation of the atmosphere. Rossby began from first principles. Solar radiation heats the surface and drives convective processes. The planet's rotation results in the formation of cellular circulation patterns, and storms transport energy from equator to poles. The general circulation, in turn, supports the formation of climate zones. His depiction of this circulation closely followed the Bergen school model (figure 2.6). Berkeley geographer John Leighly admired this approach to dynamic climatology, confiding in Rossby that it was the best way to overcome all "the loose thinking that has been done on the question of climatic changes."[70]

Rossby traced his intellectual debts to Helmholtz and placed his work at MIT squarely in the tradition of geophysical hydrodynamics. In his historical notes, Rossby emphasized the work of the Bergen school, whose graphical methods had become the principal basis for daily forecasting. He gave credit to Jacob Bjerknes, Halvor Solberg, and Tor Bergeron as the leading representatives of a team effort that established a dense network of stations in western Norway, identified and labeled the properties of air masses, and developed the concept of the cold front and warm front as elements of the ideal cyclone model.[71] Rossby is often portrayed as primarily a follower or emissary of the Bergen school, which he indeed was during his first fifteen years in America. But there was a big gap between Rossby's approach to theoretical problems and that of the Bergen school. The polar front, although important, was not central to Rossby's theory. He was not a fan of graphical methods and became furious, almost apoplectic at times, about the limitations of isobaric geometry, maintaining that meteorologists really needed to think more about hydrodynamic conservative quantities, his favorite being vorticity, a measure of local rotation in the fluid flow.[72]

Observational and theoretical work done by Rossby and Bernhard Haurwitz demonstrated that pressure changes at the surface were accompanied by offsetting pressure changes near the tropopause. This result pointed to a midlevel layer where forces were in balance and the equations of motion for the upper-air flow could be dramatically simplified. Rossby did this by using a two-dimensional model and neglecting changes due to friction, heating, water vapor, and vertical motion.[73] For Rossby, the research frontier was theoretical, and it circulated at the 500-millibar level (an altitude of about 5 kilometers). In 1940, he articulated the now famous Rossby equation for planetary waves in the upper-level winds. The iconic form—

$$c = \bar{U} - \frac{\beta L^2}{4\pi^2}$$

—relates the phase velocity c to the mean zonal flow $\bar{U}$, the Rossby parameter β, and the wavelength L. This means for short wavelengths, the phase of the waves moves eastward with the zonal flow; for very long wavelengths, the phase of the waves moves westward, or retrograde. There is also a stationary condition where the phase velocity is zero.[74]

Rossby observed that the question of the scale of atmospheric phenomena was hardly ever mentioned in the literature published between the two world wars. Why, then, are midlatitude cyclones about one thousand kilometers across and not ten times smaller or ten times larger? You often see what you are looking for. The Bergen network of surface stations in 1918—of limited areal extent but quite dense—was *designed* to intercept traveling cyclones, like the mesh of a species-specific fishing net. Rossby aimed higher. Researchers at MIT, using statistical regularities of surface weather and propagation of the upper-air waves, tried to extend forecasts from two days to five days or one week. They were able to identify larger-scale and slower-moving systems by constructing five-day moving averages of the daily synoptic surface maps, thus filtering out transient and small-scale phenomena. They also launched radiosondes into the upper air to identify patterns of temperature and pressure at or above the 5-kilometer level characteristic of the slow-moving planetary waves circling the hemisphere. The planetary waves identified in this way have linear dimensions of the order of 5,000 kilometers, with four to five ridges and troughs at any one time encircling the planet at or above the 40th parallel (figure 3.5). Because their vertical extent is about only 20 kilometers, they effectively behave as two-dimensional inertial fluid systems and may be effectively studied by experiments in differentially-heated rotating tanks.

By the early 1940s, Rossby and his group were able to construct upper-level maps of limited geographic extent. They used a brutally simplified

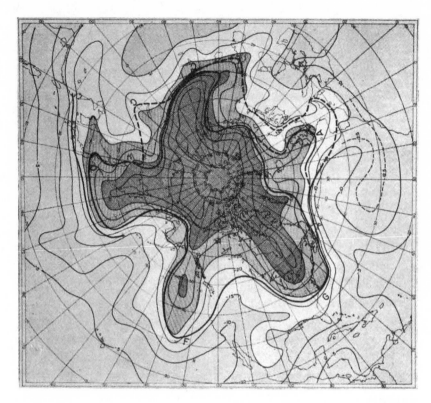

Figure 3.5
Planetary wave pattern, circumpolar view (compare with dishpan experiment in figure 3.1). *Source:* Rossby, "Current Problems in Meteorology," 20.

model, bearing some similarity to the observed motions of the atmosphere,
to analyze the wave motions revealed by those maps. They found that stationary long waves preferentially form in the lee of large mountain ranges
and over the western half of the oceans. They then analyzed how these
waves begin to move, forward or retrograde, and how shorter, traveling
disturbances pass through them triggering storms along their paths. Such
attention to the different scales of atmospheric flow patterns stimulated a
new and revolutionary focus on the energy spectrum of atmospheric
motions and how the upper air both triggers and provides support for fair
and foul weather at the surface. Sinking motion aloft leads by and large to
a disappearance of the clouds, while rising motion supports the formation
of clouds and precipitation.

Rossby was an imaginative physicist who expressed his ideas mathematically in the simplest, most intuitive manner. According to Bergeron, he had a knack for "ruthlessly simplifying a problem until it becomes solvable." This was evident in his use of rotating-tank experiments, his studies of turbulence, and his use of isentropic analysis. It was spectacularly on display in his analysis of large-scale horizontal winds and planetary waves. He wrote equations of motion for a two-dimensional plane and never, in any of his writings, used spherical coordinates. This formulation was fortunate because it allowed access to his ideas by a wide range of meteorologists and mathematicians with different levels of training. Rossby's dynamic theory of large-scale planetary flow patterns explained their movements and their importance in shaping surface weather conditions. It constituted an approximate solution for the upper-air flow, but it did not untie the Gordian knot itself. Looking to the future, the research frontier lay in improving understandings of Earth's heat budget and the stability of upper-air dynamical processes. Better models of the general circulation of the atmosphere would in turn lead to better understanding of climate and better long-range forecasts. In addition to these issues, the state of meteorological science in 1940 included the widespread use of radiosondes and isentropic charts reaching into the stratosphere, studies of fog and low clouds of importance to the airlines, and important research on evapotranspiration, ice crystal nucleation, atmospheric electricity, ozone, Arctic air masses, and ocean evaporation. Rossby had the grand vision of how it was all to be orchestrated. Although Rossby is known primarily as a theorist, he also took observations, mastered weather map analysis, and issued forecasts. Charney once called him a "compleat" synoptic and dynamic meteorologist, always the center of attention at a map discussion, in a conference, or at a planning meeting.[75]

Rossby wanted to leave the Washington bureaucracy behind, but as of June 1940, he was still employed by the Weather Bureau, deeply involved in policy decisions, and seeking to influence the national scene. He reported to MIT president Compton that Jacob Bjerknes, stranded in the United States by the outbreak of war in Europe and the Nazi occupation of Norway, had just accepted a position at the new University of California in Los Angeles and needed an assistant with theoretical training with close connections to MIT. Jorgen Holmboe, an assistant professor at MIT who had studied with Bjerknes, fit the bill. Rossby also offered Weather Bureau funding for "a small cooperative project" to study typical frontal and

air mass patterns over the Pacific. Rossby's larger goals for the Weather
Bureau involved tripling the number of Weather Bureau employees sent to
university, setting up an experimental forecast unit using the new techniques developed at MIT, establishing an inviolable research budget, ratio-
nalizing the pay grades, and identifying his successor as head of research.
Ever the talent scout, Rossby had identified as a worthy assistant chief Dr.
Charles F. Sarle of the Bureau of Agricultural Statistics—an excellent diplo-
mat, not overawed by government regulations, and with plenty of drive: "It
is quite clear that Dr. Sarle is . . . the only man that can really stir the
weather bureau and create a semblance of efficiency and progressiveness."[76]
Rossby wrote all this from Chicago, where plans for a new meteorology
department were in the works.

The Chicago School of Meteorology and the War Years

In 1940, Horace Byers and MIT graduate student Victor Starr were on assignment, teaching a three-month course for Weather Bureau employees in
Chicago. During a visit from Rossby, they decided to approach the University of Chicago about establishing a new institute of meteorology. Byers
made the phone call, and he and Starr presented a seminar on isentropic
analysis to the physics department. It was well received. Soon after, they
were in business, with funds provided by Sewel Avery, chair of the board of
U.S. Gypsum. Rossby was named chair of the Institute of Meteorology at
the University of Chicago, on leave for a year to remain with the Weather
Bureau. Byers ran the program, supported by Starr, forecaster Vincent Oliver, chemist Oliver Wulf, two statisticians, and Wexler, who had taken a
one-year leave from the Weather Bureau.[77] The next year, Rossby arrived
and began to build up the Chicago program into a powerful research school,
even as the United States plunged into world war.

World War II—fought in the air, at sea, and on land in theaters as diverse
climatically as the Sahara Desert, the jungles of Burma, the Pacific Ocean,
and the Greenland ice sheet—mobilized the meteorological talent of the
nation and its allies. Under Rossby's leadership and with input from the
army, navy, and Weather Bureau, thousands of weather officers received
their training directly through the military or at the five institutions offering advanced degrees in meteorology—MIT, Caltech, New York University,
UCLA, and Chicago. Rossby and his students or associates had started all
but one of them.[78] The war in the air turned meteorology into a key

strategic science, resulting in worldwide weather forecasting support for a large number of critical operations, the organization of the first circumpolar network of regular daily radiosonde launches, and the training of thousands of new weather officers.

The weather cadet training program got into high gear early in 1940 after Sverre Petterssen informed Captain Arthur (Merri) Merewether, chief of the U.S. Army Air Corps, that the German Luftwaffe had about 2,700 trained meteorologists; the United States at the time had about thirty. Rossby leveraged this into a nationwide effort. Most of the army students attended the Technical Training Command School in Grand Rapids, Michigan, directed by Wexler; the navy program was coordinated in Annapolis. A number of colleges also ran preparatory courses in physics and mathematics, and many meteorologists were directed to Fort Monmouth, New Jersey, for advanced courses in electronics, which was becoming more pervasive in remote sensing. About 7,000 weather officers were trained in the army program alone. All contributed to the war effort, but only a small percentage remained professionally involved in meteorology; most of those who made their mark and remained in the field either served as instructors in the wartime program or had been trained in the universities.[79] To coordinate efforts, Rossby established the University Meteorology Committee, whose members included the chairs of meteorological departments, directors of military training units, and top military officers. It provided a model of cooperation among institutions with long-lasting effects. Byers, who represented Chicago on this committee, became the first chair of a somewhat analogous group established for a different purpose in 1958, the University Committee for Atmospheric Research (chapter 5).

Another of Rossby's wartime initiatives was the Institute of Tropical Meteorology, established at the University of Puerto Rico in 1943 by the navy and the University of Chicago. It was tasked with understanding the dynamics of weather and preparing forecasts for areas far removed from the influence of the polar front. Clarence Palmer of New Zealand served as the first director, supported by hurricane scientist Gordon Dunn and instrument expert John C. Bellamy. The second director, Herbert Riehl, made his mark there training forecasters and developing models of so-called easterly waves that on occasion develop into tropical storms and regularly form into organized bands of intense convection. This work exploded the myth that convection occurs randomly in the tropics or forms only because of ice crystals aloft. Riehl went on to a distinguished career in tropical meteorology.[80]

During World War II, the U.S. Army Air Forces Air Transport Command operated a global network, establishing airfields, communication centers, and weather-forecasting facilities for the war effort. Due to the adoption of a standard international code, hundreds of thousands of observations taken worldwide were pouring in over the teletype every hour. To deal with the volume, Bellamy developed automatic plotting procedures. Military pilots routinely took risks that are unacceptable in civilian aviation. They raided enemy targets and risked difficult terrain, flying over the Himalayas in support of the Burma campaign, and distance, with routine transoceanic routes over the Atlantic and Pacific. Petterssen's role was of central importance in the European theater. He resigned from his MIT position to work in England to prepare forecasts for bombing raids over Germany and participate in the critical invasion-day forecasts for Anzio and Normandy. He was the only Bergen-trained meteorologist involved in D-Day.[81] After the war, the Provisional International Civil Aviation Organization took charge of the operation of safe commercial routes, with the main focus on transatlantic traffic. Petterssen recounted his experiences as head of their meteorology program and described how he solicited international cooperation to add thirteen floating observation posts serviced by thirty ships to the North Atlantic. This system was made a permanent part of the United Nations organization in 1947.[82]

Aircraft design was also undergoing rapid change. The Boeing Stratoliner was built in 1940 with a pressurized cabin for "over-weather" flights at up to 20,000 feet, far above the conditions below. For the Stratoliners, the main weather concerns were conditions near the ground at takeoff and landing and clear air turbulence and jet stream winds during the flight.[83] Jet aircraft made their first appearance in World War II, with the German Messerschmitt being, by far, the most common. Boeing's B-47 Stratojet bomber and B-52 Stratofortress, named for the years of their first flights in 1947 and 1952, were purchased in quantity by the U.S. Air Force. The British de Havilland Comet, the first jet airliner, heralded a new era in commercial aviation when it was introduced in 1952.[84] Such changes in aviation served to refocus meteorologists' attention on the upper atmosphere.

Rossby was an expert consultant to both the Office of the Secretary of War and the commanding general of the U.S. Army Air Forces. He was chair

of the University Meteorology Committee and head of the Institute of
Meteorology at the University of Chicago. He had the ear of the army's top
weather officer, Donald N. Yates, who had studied at West Point and Caltech
and served on General Dwight D. Eisenhower's staff in Europe. Rossby sent
Yates a seven-page memo outlining his vision of the coordination of postwar research and service. He wanted to unify operational weather activities
inside the war department and clarify the relationship between the air
weather service, which had grown by leaps and bounds, and other governmental and private meteorological institutions and organizations: "The
need for intimate cooperation between meteorological research institutions
and the operating weather organizations of the government is an urgent
one. Unless a formula is found for such cooperation, I fear greatly that the
operating weather services may continue to grow in size but reach stagnation as far as technical improvements are concerned."[85] Rossby made it his
policy to avoid stagnation. He found strength in diversity, opining that
research, training, and operational forecasting needed to go hand in hand.
As a case in point, Rossby noted that England and France, both victorious
in World War I, had markedly strengthened centralized government
control over all meteorological activities. The French weather services
were centralized into one single organization and, in Rossby's opinion,
had not given adequate consideration to the need of free research in meteorology at independent institutions. As a result, French contributions
to meteorology languished and were, on the whole, insignificant. In England, the weather service became a quasi-military organization under the
air ministry, again at the expense of research and training. The opposite
happened in Germany and Austria. The Treaty of Versailles weakened government power, and research in meteorology flourished in about a dozen
academic institutions, although not on a large scale. In the Scandinavian
countries, untouched by the war, meteorology reached world prominence
at institutions that coordinated with governmental needs but maintained
their academic independence. Rossby's bottom line was that although coordination is necessary, centralized control of research and development
activities would likely lead to "a drying up of the springs from which new
ideas flow."[86]

During World War II, the military meteorological services in Russia, Germany, and Great Britain were led by experienced professional civilian meteorologists. By contrast, U.S. weather officers were primarily soldiers and

only secondarily meteorologists, most with only nine months of training and no research experience. Rossby and his civilian cadre had served mainly as teachers or consultants, with little say in policy matters. This was an acceptable compromise for the war emergency but would not be sustainable in the long run. According to Rossby, the military system tended to reward aggressiveness and leadership skills over scholarship and was channeling the weakest students into important administrative positions. The combination of weak technical knowledge and high rank was likely to become more and more embarrassing, both to them and to the army weather service. He worried that most were not prepared to answer technical questions in meteorology or even know where to look to find them. Some held their jobs "by virtue of a pleasant personality and good salesmanship rather than by virtue of genuine and versatile technical ability." Although some could prepare viable forecasts, he feared the large and growing number of "eager amateurs."[87]

Rossby wanted to find a better way of promoting cooperation between the military and universities by involving faculty in military research rather than maintaining the status quo in which military units were peripheral. His was an expanded vision of dynamic and inseparable relations between research and service, research and training, and civilian and military sectors. With longer-range aircraft flying higher and faster than ever before, the need for an all-weather flight program staffed by qualified experts in atmospheric physics was becoming more and more acute. Rossby suggested a continuous professional training program, selective in admissions, spanning two and a half to three years. He thought the air weather service was becoming overspecialized by offering too many unrelated training courses and suggested a broader approach to training that would include an electronics course for the officers as well. The new technology of radar was promising (chapter 4), but Rossby thought its development had been stifled by security restrictions that kept it largely in the hands of junior military personnel.[88]

Early in 1944, Rossby was elected to a two-year term as president of the American Meteorological Society. By all accounts, he "plunged into the job" with characteristic zeal, announcing a major reorganization plan to place the society on a firmer scientific footing. Rossby established a formal office of the executive secretary headquartered in Boston, instituted a new category of membership requiring advance training in meteorology, reemphasized professional standards and ethics, expanded opportunities for the employment of meteorologists in the private sector, persuaded the Weather Bureau to make weather information over teletype facilities widely available and free of charge, and launched a flagship publication, the *Journal of Meteorology*.[89]

In an article in the first issue of the new journal, Rossby replaced purely empirical forecasting rules with a more complete theoretical explanation of horizontal motion in the atmosphere. In the second issue, he developed the important concept of group velocity, in which energy propagates or is dispersed through a group of oceanic and atmospheric waves of different frequencies. Drawing by analogy on the work of Harald U. Sverdrup for the propagation of ocean surface waves, Rossby was able to explain the propagation of energy as it passed through the atmosphere, faster than the waves themselves. Rossby described two synoptic cases—one in which energy of atmospheric planetary waves propagates rapidly eastward over great distances and one in which the phase velocity is negative. This meant that the earth's atmosphere was interconnected over great distances and that influences propagated faster than even the strongest winds, making it impossible to isolate certain regions and analyze them as if they were mechanically or thermally closed systems.[90] Rossby had, once again, assembled an outstanding group around him, including graduate students Victor Starr, Morris Neiburger, and Dave Fultz; meteorologist Erik Palmén from Finland; and, fresh out of UCLA, Jule Charney, who spent the 1946 academic year at Chicago. Together Rossby and Charney were able to explore the basic concepts of planetary waves, the general circulation, and the jet stream.[91] Their relationship allows a new look at Rossby through Charney's collection of letters and his memories later in life.

Charney and Rossby

In the spring of 1942, Jule Charney read Rossby's 1939 paper on planetary waves and presented a seminar on it at UCLA that generated "almost no discussion" at the time. His graduate program simply did not have a critical mass of theorists. When asked what guidance he had received directly from his professors Jacob Bjerknes and Jorgen Holmboe, Charney replied bluntly, "None whatever." When asked, "Did you seek guidance?," Charney replied simply, "No," adding later, "I got no assistance." Charney, an autodidact, then began a more careful theoretical study of the stability of long waves that led to his thesis. His "neat" derivation" agreed with Rossby's formula and provided a mathematical "vindication" of Jacob Bjerknes's more qualitative ideas about the deepening of troughs. Charney, in his own words, had solved "The problem of the motion of the long waves in a baroclinic atmosphere . . . as a perturbation of a basic flow in which there were horizontal temperature gradients and vertical shears of the zonal wind."[92]

Jule Gregory Charney was born in 1917 in San Francisco into a family of emigrant Russian Jews from Belarus. His father worked in the garment industry, and his mother as a lady's tailor; both were well read and politically active. In 1922, the family moved to Los Angeles. Charney was precocious in math, teaching himself calculus and differential equations, which he called "duck soup." He was particularly attracted to solid geometry and mastered its axiomatic system, deriving from that a "geometric sense and some feeling of power." Charney entered UCLA in 1934 at the age of seventeen. The new Los Angeles branch of the University of California, with five to six thousand students, was just getting underway. Charney had no instruction in theoretical physics and was forced to take college algebra and calculus courses he had already mastered: "I knew all that stuff. . . . I didn't have to study at all. . . . I really had nobody to inspire me mathematically." Charney recounts how he studied electricity and magnetism without being introduced to Maxwell's equations and optics without reading Kirchoff's work on diffraction: "There was no theory! And, you know, it didn't send me."[93]

Charney graduated from UCLA magna cum laude with a bachelor's degree in 1938. Despite his poor opinion of the place, he remained there as a graduate student in mathematics and physics from 1938 to 1941. He declined an invitation from a mathematics professor to submit a good but, in Charney's opinion, not very original paper for his PhD in mathematics. Instead, he was appointed a university fellow and continued his research. In the spring of 1941, Charney heard a lecture from Holmboe on elementary hydrodynamics and began to see meteorology as "a fairly serious science." Before then, Charney confided, "I knew zero about meteorology," adding that he came to the field "not being fully convinced that wind was moving air."

War was in the air in 1941, and many of Charney's friends were either joining the army or seeking positions in defense-related industries.

Holmboe offered Charney an assistantship and invited him to take the nine-month wartime training course in meteorology while serving as his assistant. Charney sought the advice of the noted aeronautical engineer Theodore von Kármán at Caltech, who told him that the best choices were in either the aircraft industry or meteorology, with the latter being more theoretically oriented. Charney transcribed the course notes for Holmboe and, as an advanced student, helped grade the papers in the dynamic meteorology course he was taking. Charney had this to say about Holmboe: "He was very careful; he was very visual; he could only understand things geometrically. He had very little analytic ability . . . he had to see vorticity geometrically." Charney "*hated* the laboratory." He much preferred theoretical analyses of the global circulation and thought that plotting U.S. weather maps was "a total waste of time." Charney also studied radiation with Joseph Kaplan and authored the chapter on radiation for the 1945 *Handbook of Meteorology* published by McGraw-Hill. This was his only publication on radiation.[94]

In 1944, Charney began "really doodling with the equations of motion" and, before long, had derived a set of equations that, "to my great delight and amazement, turned out to be a very simple hypergeometric equation. . . . I remember going for a walk in the hills . . . and just transforming them in my head and suddenly I saw the . . . solution."[95] There was a direct link in Charney's work to the fluid dynamics of Helmholtz and Kelvin in the nineteenth century, the hydrodynamic stability problem of William Orr and Arnold Sommerfeld circa 1908, and the work of the Soviet mathematician Nikolai Kochin, transmitted into meteorology by Vilhelm Bjerknes, Halvor Solberg, Jacob Bjerknes, and Rossby. The work "certainly gave a big boost to my career. . . . It (at last) convinced me that I could do research, and it solidified my love of meteorology. Here was a field where I thought I could accomplish something." Later in life he reflected, "I don't think that anything I did in numerical weather prediction was comparable in originality."[96]

As Charney was nearing the end of his graduate program, the UCLA faculty suggested he should study in Norway with Solberg. He applied for but was not awarded an American-Scandinavian Foundation fellowship, but he won a prestigious fellowship from the National Research Council that allowed him to travel. Jule and his wife, Elinor, left Los Angeles in the summer of 1946 but were detoured on their way to Norway. Elinor headed for a family visit in St. Paul, and Charney visited Rossby in Chicago. He had met Rossby once when he lectured at UCLA: "I came to Chicago and Rossby was . . . extremely welcoming and very kind. . . . They asked me to give a seminar on my thesis, which I did. And in those days things were extremely informal."[97] Charney's seminar provided a theoretical explanation of the origin and development of extratropical cyclones. His starting points were the hydrodynamic theories of Vilhelm Bjerknes and the polar-front theory of the Bergen school. Solberg had developed a simple theoretical model of unstable waves along a sloping surface separating two atmospheric layers and Jacob Bjerknes examined divergence in the troughs of upper-air waves as the cause of cyclone intensification. The relationship between upper-air and surface weather phenomena, however, was not clear. Upper-air waves can be three times larger than surface disturbances and typically move at different speeds. Moreover, waves have crests as well as troughs, and an explanation of their stability was lacking. There were probably not more than a dozen people present. Charney presented a highly mathematical seminar lasting several hours, complete with derivation of the equations: "They sat patiently through the thing."[98]

Charney claimed that the Chicago seminar was the first time he really had a stimulating theoretical discussion of the general problem of long waves. Charney published the work in the *Journal of Meteorology*, taking up almost the whole issue; demand for reprints was so robust they were soon exhausted. His fundamental result—that the Rossby formula for the speed of propagation of long waves in a barotropic atmosphere was a viable approximation—opened the door for numerical weather prediction. The work was meteorologically interesting, mathematically elegant, theoretically sound, widely available, and widely cited. It led to Charney's personal relationship with Rossby, whom he considered to be "an intellectual godfather." He also thought this paper was "probably" his most important work, in the sense that it had been the most influential. [99] After that, Charney recalled, "Rossby was *very* friendly and we had some *very* good scientific discussions on my thesis and related matters. He emphasized very much the distinction between these long waves and the shorter waves that one sees on the surface weather map."[100] Rossby, who was thinking two-dimensionally in those days, benefitted from Charney's insights into the vertical propagation of planetary waves. Rossby loved a heuristic argument, sometimes switching sides to play devil's advocate: "You could argue with

Rossby, you could challenge him, whereas I would never dream of doing that with Bjerknes."[101]

Charney delayed his trip to Oslo and the start of his fellowship, accepting Rossby's invitation to become a research associate at Chicago for the new academic year. He lectured in the hydrodynamics course and was able to attend two important planning meetings. One was convened by John von Neumann in Princeton in August to discuss the meteorological computing project, and another was held in Chicago in December on research problems in meteorology (chapter 4). Rossby and Wexler set the agendas for those meetings. When asked by George Platzman, "Do you think the whole Chicago experience was an important one?" Charney replied, "Oh, I think I might even say it was the main formative experience of my whole professional life. . . . Rossby and I were, in a way, kindred spirits. I mean we thought very much the same way, and we had *endless* conversations . . . And he always made himself available . . . . I discovered what it meant to have intellectual rapport with another person."[102]

Between the end of the winter quarter at Chicago and his March 22 departure for Oslo, Charney squeezed in a short trip to Princeton to see his UCLA cohort Philip Thompson, the only person working full time on the meteorology project. Charney shared his first impressions in a letter to Rossby: "Right now the Princeton met[eorological] project is the ugly duckling of meteorology, but there are indications that this ugly duckling may become a swan. . . . My overall impression from talking to von Neumann, Thompson, and [Herman] Goldstine, the co-ordinator for the project, is that the computing machine will probably be able to integrate the equations of motion. What remains now is to tell the machine what the equations of motion are and to give [it] the proper data."[103]

In Oslo, Charney's office was located next to Solberg's, but the two had "essentially no scientific discussions." Charney's peer group consisted of Arnt Eliassen, Ragnar Fjørtoft, and Einar Høiland:

And, of course, V. Bjerknes was very much there. We had luncheon very often. . . . I remember talking to him [not about my work]. I found his reminiscences to be absolutely fascinating, because he had been an assistant to his father, [Carl Anton] Bjerknes who was influenced by and had influence on people like Kelvin and Helmholtz and . . . Hertz. I felt a kind of a historical connection. Helmholtz was then and remains one of my great scientific heroes . . . [and] I'm sure V. Bjerknes never lost his interest in electromagnetism.

Charney and his small circle often did not go home for their main midday meal, as was the tradition, but instead went to a little restaurant called The Valkyrjen to continue "rather beery, but very exciting scientific discussions."[104] The fellowship provided Charney with "time and tranquility to start really thinking about the problem" of filtering the equations of motion to eliminate fast-propagating gravity waves. He was wary of Lewis Fry Richardson's much earlier attempt to integrate the primitive equations and referred to his "fundamental error" of not being able to calculate the divergence in his initial tendency field.[105] In 1922, Richardson attempted to perform a numerical integration of the atmospheric equations. However admirable, it was an isolated, heroic, yet quite impractical effort requiring several years to produce a retrospective six-hour forecast over Europe by exceedingly laborious hand computations. Richardson was aware of, but chose to ignore, the warning by Max Margules in 1904 that the continuity equation could lead to great errors in forecast pressure changes caused by small, unmeasurable differences in wind speeds across a region. Richardson's forecast was unsatisfactory, with no agreement between the observed and computed state. The time step he used was too large, the data array he used was too crude and too small, and the continuity problem showed up in the very large and unrealistic 40-millibar pressure change he calculated at the surface. One thing Richardson did accomplish was to warn others away from attempting to integrate the general atmospheric equations of motion using numerical hand computations. Charney was seeking a different approach.

Rossby knew and liked everyone in the Oslo group but had his intellectual differences. He praised their passionate devotion, unparalleled scientific idealism, and familiarity with theoretical methods and mathematical techniques. He warned Charney, however, of the "utter sterility" in their problem formulation and selection of topics for study. Rossby thought that their founder and his own mentor, Vilhelm Bjerknes, "the last of the classical theoretical physicists," wielded too much influence and that his deductive and deterministic approach was "uninfluenced by all that has happened to our evaluation of science since 1900." The older program involved solving the hydrodynamic equations in all their complexity to learn something about the atmosphere. The "more modern point of view," which "has not in the slightest penetrated the Oslo atmosphere," reduced these equations to subordinate tools to study problems suggested by the atmosphere itself.

Rossby also rued the "utter lack of contact between synopticians and theo-reticians" in Olso, Fjørtoft being the only exception.[106]

Charney was convinced that weather forecasting is a computing prob-lem and that its solution requires "one highly intelligent machine and a few mathematico-meteorological oilers" (see chapter 4). He was also con-vinced that he knew how to integrate the equations of motion. In a letter to Thompson, he recommended Eliassen and Fjørtoft for the Princeton project, both trained by Vilhelm Bjerknes, as the two most promising Nor-wegians he knew. He also recommended himself, offering to chip in his "two øre" for a while. He had rejected job offers at UCLA and Chicago and hoped to work in Princeton:

Rossby expects me to go back to Chicago and [Jacob] Bjerknes to California, but I can't see either of these deals at the moment. Rossby is scheduled to remain in Swe-den for the next two years, Starr has left for MIT, and while Palmén is a hell of a swell guy, I don't feature knocking around all alone with him and Byers—both synoptic men. As to California—it has a fine climate.

Charney had developed a set of filtered prognostic equations for the initial three-dimensional pressure field and thought it should soon be possible "to start turning the crank." Wexler had already invited him to join the Princ-·eton project, and he recalled similar encouragement from von Neumann. He asked Thompson, "What is the setup now? . . . What the hell's happen-ing in Princeton?"[107]

Rossby Moves to Sweden

In January 1946, Professors Harald Norinder of Uppsala and Hans Wil-helmsson Ahlmann of Stockholm made an official recommendation that Rossby be hired to strengthen meteorological training and research in Swe-den. Tor Bergeron, one of the strongest apostles of the Bergen school and a promoter of the intuitive and artistic craft of synoptic analysis, also received a chair, at Uppsala. Rossby visited Sweden that month and met with the minister of education, Tage Erlander, who offered him a position at Stock-holms Högskola. Rossby accepted the offer and left the United States for Stockholm on September 19, 1947.[108] He presented lectures to the Swedish weather service and military representatives on the theory of planetary waves, inertial gravity waves, group velocity, and the general circulation of the atmosphere. Bert Bolin, then with the Swedish air force, attended these

lectures and felt they provided a strong foundation for further collaboration among interested groups.

In Sweden, Rossby founded the international geophysical journal
Tellus (Latin for "Earth"), established the International Meteorological
Institute, built another world-class research group, added numerical
weather prediction and atmospheric chemistry to his portfolio, and continued a lifelong relationship with Woods Hole while maintaining both a residence and a research professorship at Chicago through 1951. Rossby
confided to Jacob Bjerknes, and perhaps to others, that he had often seriously thought of "abandoning all his empire building in America" and
returning to the old country.[109] In the late 1930s he had contemplated
returning to Sweden to work closely on theoretical issues with the
oceanographer Vagn Walfrid Ekman in his "ivory tower" at the University
of Lund, about 60 kilometers east of Copenhagen. Now, approaching age
fifty, a convergence of influences had brought Rossby home.[110] Europe
needed him, Sweden needed him, and the Swedish weather service
needed him. He loved his native country, and he felt honored at being
invited to return. He had family in Sweden, and his brother was involved
in high-level computing there. He could be more European and cosmopolitan, attend lectures, plays, and operas and indulge in a wide range of
extracurricular reading.[111] In Stockholm, Rossby was back in his element
(figure 3.6).

Rossby no longer found the meteorological scene in the United States to
be very stimulating. He had experienced administrative difficulties with the
military, the Weather Buearu, and at Chicago, and the focus of the American Meteorological Society was on membership, not research. He was ready
for the next set of opportunities and challenges. He maintained a visiting
professorship at Chicago and left the management of the Institute in Byer's
qualified hands. The U.S. military, Weather Bureau, and National Academy
of Sciences still needed his services, and Woods Hole welcomed his presence, but he could keep many irons in the fire through correspondence and
plenty of international travel.

Rossby actively pursued numerical weather prediction in Sweden in an
era in which there was no Swedish word for *digital computer*. In 1947, he
and his younger brother, Åke, head of code breaking at the National Defense
Radio Establishment, attended a high-level government meeting to discuss
"super calculators." Computer applications in meteorology were on the

Figure 3.6
Carl-Gustaf Rossby in Stockholm, October 6, 1954. *Source:* Wexler Family Papers.

agenda.[112] During the next several years, Rossby and his network of students and associates circulated freely between Stockholm and Princeton, developing strategies and experimental procedures for numerical weather prediction. Simultaneously, Sweden was building the BESK (Binary Electronic Sequence Calculator), which, when it became operational, aimed to be the fastest computer in the world (chapter 4). Rossby convened several international conferences on numerical prediction, and Bert Bolin took charge of developing a mathematical model for the new machine. In Bolin's words, "Purely barotropic processes define the stage upon which the thermal [baroclinic] play is performed."[113]

Rossby, fully engaged in two countries and two continents, maintained a large correspondence network of colleagues in the United States and welcomed visitors, including Charney, Jerome Namias, Petterssen, and Riehl, to Sweden.[114] This produced synergy that generated new ideas, attracted strong students, and had a positive effect on operational forecasters and government bureaucrats. He was engaged in projects, meetings, and travel that promised to reestablish personal contacts between and among scientists in Austria, Germany, Scandinavia, Eastern Europe, and beyond. From 1951-1955 the Rossbys resided at Stockholm Observatory, situated on *Observatoriekullen*, a prominent hill in the center of the city. The Geographical Institute at Stockholms Högskola had taken occupancy of the building in 1934, where they conducted polar research and made weather observations, continuing a series dating from the 1750s. The transit of Venus had been observed there in 1761. It was an iconic place for Rossby to think about global issues, build his research team, and edit *Tellus*.[115]

Rossby wanted to make *Tellus* "a truly international organ for international cooperation in a world which suffers from all kinds of iron curtains, a true mirror of the things that occupy geophysicists in our days, and a bridge between the individual geophysical sciences."[116] The first issue appeared in March 1949. As promised, it included articles from fields of geophysics other than meteorology, including Hans Pettersson's study "The Geochronology of the Deep Ocean Bed," a study of solar flares, and Hans Ahlmann's note on the upcoming 1949–52 Norwegian-Swedish-British Antarctic Expedition. The bulk of the contributions, however, were meteorological, with original papers by Erik Palmén, Tor Bergeron, and Rossby, who, using results developed together with his student T. C. Yeh, wrote an important article on the dispersion of planetary waves in a barotropic atmosphere. New patterns were emerging from theory that pointed to practical applications. An article in the second number of *Tellus* by Charney and Eliassen, "A Numerical Method for Predicting the Perturbations of the Middle Latitude Westerlies," excited Rossby so much he wanted to see it developed into an applied technique for forecasters. He called for a "systematic test and extension" of such methods to "get rid of the horrible subjectivity which still characterizes all, or almost all forecast efforts." He felt the revolution coming and wrote to Platzman, "I must confess that I have an extremely strong feeling that we are standing at the threshold of a new era in applied meteorology and that we must push this line to the point where it can be put in general operation." *Tellus* published Byers's article on the structure and dynamics of thunderstorms, Riehl's paper on the role of the tropics in the general circulation of the atmosphere, and Wexler's analysis of annual and diurnal temperature variations in the upper atmosphere. After the appearance of a landmark paper by Jule Charney, Ragnar Fjørtoft, and John von Neumann on numerical integration of the barotropic vorticity equation, Charney wrote to Rossby, "[*Tellus*] is certainly maintaining its reputation as the liveliest and meatiest journal in meteorology." It was on a firm foundation.[117]

Rossby convened numerous meetings, including two on postglacial climatic fluctuations and another on problems related to cyclones, jet stream structure, and the general circulation that brought scientists together from over a dozen different nations in informal and egalitarian settings. The International Meteorological Institute formally opened in 1955 with support from the Swedish Institute for Cultural Exchange with Foreign Countries and funding from the Swedish government and the U.S. Office of Naval Research. Alluding to the indivisibility of the atmosphere and the pressing needs and opportunities in the postwar era, Rossby called for additional organizations to support large-scale cooperative research projects on an international basis. The "Rossby Institute" provided a model for the U.S. National Center for Atmospheric Research and the international exchanges of the 1960s (see chapter 5).[118]

The typical situation, at least in Europe, both before and after the war, consisted of a lone university professor with one or two assistants attempting to keep up with and contribute to the advancement of science. While the scale of operations was much larger in the United States, the focus was on domestic and military weather services, with overburdened administrators and forecasters struggling to keep up with routine tasks. Rossby cited as a notable exception the "brilliant achievements" of the Bergen school, assembled under the guidance of Vilhelm Bjerknes, that promoted teamwork and innovative research while providing valuable daily forecasts. Rossby sought to place meteorological science ahead of politics. He thoroughly enjoyed building new research teams and engaging in discussions of new ideas and theories, but he intensely disliked the red tape and bureaucratic restrictions imposed by national service, including defense work. He compared research by national establishments to Olympic teams competing against one another for medals. They were separated by culture,

language, travel restrictions, and often pride. He sought, through cooperative research on issues of mutual interest, to heal many of the postwar rifts
that had developed. Although he had been granted a top-level security
clearance for his wartime work, his focus was on diversity and international
collaboration as essential elements of progress. He noted the rise of English
as an auxiliary language by the talented Chinese and Japanese researchers
he knew and thought that such communication strategies were essential
everywhere. He felt deep gratitude for the "escape from pomposity" and the
intellectual opportunities he had experienced in his work in America.
Bringing this ambience to Europe was more of a challenge: "The European
chinashop is a delicate one and must be handled with a great deal of care if
we want to avoid serious damage to its antiques." Ruing the isolation and
irrelevance of some of his colleagues and looking to America and the Soviet
Union, Rossby wrote, "On both sides of us basic science has been organized
much more completely into service for the community." He hoped that the
United Nations Educational, Scientific and Cultural Organization and the
World Meteorological Organization, working together, would sponsor
international scientific laboratories and independent research teams that
could work across national boundaries and facilitate the daily collection
and compilation of weather observations from national services.[119]

On the Chemical Climate

"A problem solved is a dead problem," Rossby often remarked, and in the
union of dynamic meteorology and environmental chemistry, he found a
host of new challenges. Beginning in 1952, he convened a series of conferences in Stockholm to investigate the linkages between local processes and
global geochemical cycles. The participants called their work "climatological studies in atmospheric chemistry" or simply "chemical climatology."
They issued an international call for observational and research plans to
cover much of the world.[120] Rossby was interested in investigating the role
played by the atmosphere in the circulation of various chemical substances.
The geographical and geophysical tradition in Sweden inspired Rossby to
initiate a network of stations to observe the chemistry of the air and rainwater. This represents the beginning of the research field of atmospheric,
particularly tropospheric, chemistry. His goal was a coordinated and cooperative network of chemical stations spanning the Northern Hemisphere

that would collect data on the geographic distribution of various chemical species—their sources, sinks, and concentrations—as observed directly in the air, the soil, or precipitation. Information on sources and sinks over land and sea was also needed. The data collected this way were displayed and analyzed on synoptic weather charts in an attempt to understand their dynamics.

Rossby's specialty was studying the role of the atmosphere as a carrier and distributing agent for matter such as maritime salts, dust, industrial pollutants, and radioactive debris. He was aware of the increasing demands on meteorologists to be able to calculate the future path (or trajectory) of an air parcel containing such materials. This problem was much more challenging than numerical forecasting at the 500-millibar level, because actual three-dimensional winds are sought. He thought it likely that aerosols released in a limited area might accumulate in more stable layers, such as temperature inversions, or flow in concentrated streaks due to strong wind shears, such as in the jet stream. The regular network of synoptic observations and regular forecasting methods, even the new numerical ones, did not reveal such details of the wind patterns. But a new (classified) U.S. Air Force program promised to do so by launching neutral-density balloons that move with the winds along constant pressure surfaces and tracking them with radio direction finders. Rossby hoped that the World Meteorological Organization might duplicate this program for the benefit of civil society and for research scientists hoping to understand trajectories and turbulent mixing in the atmosphere.

Rossby also focused his attention on the chemistry, transport, and diffusion of large-scale and permanent sources of aerosols, such as sea salts. Wind-driven white caps on the sea transfer droplets of water containing salt particles into the atmosphere, which, at some later time, act as condensation nuclei for cloud and rain formation. To illustrate this effect, Rossby's associates constructed chemical-climatological maps of precipitation in Scandinavia showing the amount, salt concentrations, and distance inland from the coast. The network of stations was sponsored by the Royal College of Agriculture and the International Institute of Meteorology in Stockholm. Scientists there conducted chemical analyses that revealed for the first time the phenomena of acid rain and acid snow, with hydrogen ion concentrations (pH) of precipitation measured at or below 4.5, low enough to have a negative influence on fish reproduction. Swedish studies revealed "with good probability" that an airborne contribution was coming from sulfur emitted by British and German industry. Other topics of interest included Los Angeles air pollution and the trajectories of artificial radioactivity from bomb debris.[121]

In 1956, Harry Wexler visited Sweden to discuss plans for atmospheric and geochemical observations during the International Geophysical Year. With the U.S. Weather Bureau taking the lead, scientists were planning chemical investigations of the atmosphere, oceans, and rivers. They also were developing new techniques of drilling and analyzing glacial ice and ocean sediments. Stations in the Arctic, Antarctica, and Hawaii would measure precipitation, oxygen, ozone, particulates, radioactivity, and trace gas concentrations, including carbon dioxide using a new infrared measuring technique. The British scientist Guy Stewart Callendar had raised important questions about the natural exchange of carbon dioxide between the sea and the atmosphere, its role in Earth's heat budget, and its unnatural (that is to say, anthropogenic) increase due to fossil fuel burning and deforestation. Scandinavian researchers, inspired by Finnish chemist Kurt Buch, had constructed networks for measuring variations in carbon dioxide as possible air mass tracers and discussed the desirability of remote monitoring sites in the Sahara Desert or on a Hawaiian mountain.[122]

Bert Bolin called his associations with Rossby over a ten-year period "decisive for my career" (figure 3.7).[123] Rossby involved him in an international network of excellent researchers in Stockholm that created an "exceedingly stimulating" intellectual environment. Bolin, in later life a founder of the Intergovernmental Panel on Climate Change, brought some needed clarity to the issue of atmosphere-ocean interactions involving carbon dioxide. He and Erik Eriksson developed a detailed chemical and dynamical model of the short-term exchange of carbon dioxide between the atmosphere and the upper mixed layer of the ocean. They explained the mechanisms of seawater buffering and exchange in clear terms and emphasized what it meant. Their model indicated that the surface ocean layer must have taken up less than 10 percent of fossil fuel emissions and was acting as a "bottleneck" in the transport of carbon dioxide to the deep sea. They concluded, in agreement with Callendar, that if industrial production indeed climbed exponentially as projected, atmospheric carbon dioxide would probably rise 25 percent or more by the end of the century. The community of geoscientists was beginning to realize that Earth was

Figure 3.7
Carl-Gustaf Rossby and Bert Bolin at the International Union of Geodesy and Geophysics Rome Assembly, September 19, 1954. *Source:* Wexler Family Papers.

vulnerable to anthropogenic pollution and that humanity could not rely
on the oceans to absorb all the emissions of fossil fuels.[124] Rossby agreed,
writing that "we already have some observational evidence and experience
that indicates the possibility of unintentional or intentional human interference with . . . the climate at the earth's surface. One indication of that
kind [is] the effect on the mean temperature of the atmosphere of the
increased carbon dioxide content caused by the continuously increasing
consumption of fossil fuel."[125]

Rossby supported improved understanding of cloud physics, but was
largely critical of recent attempts to control atmospheric processes directly.
Some scientists, notably Irving Langmuir, had conducted experiments to
interfere with the natural water cycle to stimulate the release of precipitation locally or regionally; others speculated that the use of surface films
might reduce evaporation from reservoirs and snowfields. Historically,
"rainmakers" were considered charlatans. A new generation of scientific
and commercial rainmakers did not fare much better, especially when their
claims outran existing knowledge of cloud systems and their internal

microphysical processes. Rossby thought that most practitioners naively misunderstood the natural forces involved in the formation of precipitation and operated under crude assumptions within a very narrow window of influence. Although they could intervene in clouds, with chemicals such as dry ice and silver iodide, they could not control them. Some were advocating research on large-scale experiments to change the albedo of the earth, but Rossby warned that we cannot forecast with any precision the probable consequences of such a planetary-scale intervention.[126]

Rossby's Sudden Demise

Rossby had a weak heart, a condition that developed in his youth after a bout with rheumatic fever. On one occasion in 1952, he had been feeling overworked and out of sorts but still arose early in the morning. He wrote to Charney: "This is written at my desk at 5:30 a.m. I have seen the sun rise over the City, the sky is clear, my family is sleeping, and I can hear Tommy's breathing in the next room. All is well with the world. I am beginning to feel rested and pray to God the time will come when I will feel like going out to sock somebody on the jaw scientifically speaking, on general principles. You can see I am recovering."[127]

As the International Geophysical Year got underway, Rossby was busy planning the September 1957 Toronto meeting of the International Association of Meteorology and Atmospheric Physics. He had asked Charney to chair a full-day session on numerical forecasting and dynamic meteorology and present an introductory survey lecture.[128] Rossby never got to Toronto. On August 19, 1957, during a conference in Stockholm, he was found collapsed at his desk, the victim of a heart attack. His sudden demise came as a great shock because Rossby had much more to give. The next morning, Charney penned a eulogy:

The death of Carl-Gustaf Rossby is an irreparable loss to the whole world of meteorology, but to Americans the loss will be felt especially keenly because Professor Rossby, with his scientific genius, warmth of personality and organizational ability, was in a literal sense the father of modern American meteorology. He founded the first department of meteorology in the United States at MIT and later the meteorology department at the University of Chicago; he was adviser to the Weather Bureau, the military weather services and spiritual godfather to two generations of meteorologists. There are few meteorologists in America who have not warmed themselves in the sun of his vivid personality and few who have not fallen into his intellectual debt.[129]

Charney remarked on that sad occasion: "Without Rossby, my world has become dim."[130] Bolin, twenty-seven years Rossby's junior and his closest associate in his later years in Sweden, wrote the following in memoriam:

Professor Rossby will be especially remembered for his generosity, friendliness and sparkling humor. He took great interest in his students and often expressed the idea that you have to consider every one of them as a special project to be developed according to the personality involved. His enthusiasm and optimism kept him spiritually young and it is characteristic that until the end, it was a group of young people he gathered around him. Rossby had the simplicity of the great. His life was his work. He died during a conference making great plans for the future. Carl-Gustaf Rossby is dead, but his name and his spirit will be carried on by friends, students, and colleagues all over the world.[131]

In the memorial volume, Bergeron evaluated Rossby's accomplishments through the lenses of ongoing improvements to observations (O), tools (T), and models (M). Rossby took a leading role in all of these. He accomplished this instrumentally with rotating-tank experiments, aircraft soundings, and use of radiosondes; programmatically through the introduction of air mass, frontal, and isentropic analysis in the United States; and theoretically by introducing new concepts regarding the conservation of vorticity, planetary long waves, group velocity, jet streams, the general circulation of the atmosphere and oceans, turbulence, and chemical climatology.

Rossby worked in meteorology in an era of the ascendance of commercial and military aviation, and this, like echoes on radar screens, brought the science to the attention of the top levels of government. His schools of thought at MIT, Chicago, and Stockholm developed new mathematical techniques for data assimilation, analysis, and forecasting. His simplified model of large-scale dynamics treated the atmosphere in two dimensions, focusing on the 500-millibar level and above. It resulted in equations for idealized upper-air flow amenable to solution using the new digital computers. In his Stockholm period, he promoted the study of chemical meteorology and climatology. Near the end of his life, Rossby was deeply involved in an attempt to establish a World Meteorological Organization research project on the meteorology in the arid zone and the possible creation of a department of meteorology at the American University in Beirut, Lebanon.[132] His final essay, "Current Problems in Meteorology," written in 1956, examined meteorology's global frontiers—planetary-scale energy balance and ocean circulation, carbon dioxide and climate, the general circulation of the atmosphere, weather forecasting using new technologies, the transport and distribution of aerosols and chemicals, and the possibility of artificial control of atmospheric processes. It was published posthumously in English in 1959.[133]

Rossby influenced everyone who knew him—students, colleagues, and those outside the discipline. He built the most significant organizations of his era, started new research schools, responded unselfishly and creatively to the war emergency, and afterward, worked to reunite estranged colleagues and chart new directions for atmospheric science.[134] His work took a global, environmental turn in the mid-1950s. He fostered new conversations among geoscientists of all stripes—oceanographers, geographers, and geologists. He was interested in climatic change and variability on all time scales, including the grand cycles of ice ages and interglacial epochs. He spoke increasingly of the atmosphere as a milieu that directly influences all of human experience and warned of the increasing stress that pollution was placing on it. He stood in awe of its dynamical and chemical complexity and called for an attitude of respect "for the planet on which we live." He eagerly anticipated the coming breakthrough, "a grand era in meteorology" when artificial satellites can view the atmosphere from above. "Right now, we are like crabs on the ocean floor," he said. "What we need is a view from a satellite. Only from a satellite can we see the planetary waves."[135] His best student, Harry Wexler, was in the process of developing such capacity.

4 Wexler

One of Harry Wexler's most significant trips was a thirty-minute walk he took across Cambridge, Massachusetts, in 1932 to see Professor Carl-Gustaf Rossby. It started at Harvard University, where he had just received his undergraduate degree, magna cum laude, in mathematics with an honors thesis on fluid mechanics. He was ambivalent about his next steps and had no clear idea about what direction to take or what discipline to pursue, perhaps a field that combined his mathematical skills with a more applied physical science. His childhood friend Jerome Namias had made the introduction, and the bonding was instantaneous. On the occasion of his twenty-fifth college reunion, Wexler recalled, "A five-minute talk was enough—even if Rossby had been a professor of classical Abyssinian philology I would have wanted to study under him. Thus the torch for research meteorology, once lit, has continued to glow."[1] In fact, Wexler's torch burned brightly for the next thirty years.

Grand-student of Vilhelm Bjerknes and student of Rossby, Harry Wexler was one of the most influential meteorologists of the mid-twentieth century (figure 4.1). His influence, as head of research in the U.S. Weather Bureau, touched every aspect of weather and climate science, all of the relevant technologies, and important aspects of the related geophysical sciences—oceanography, glaciology, and ionospheric physics. He was a master of chemical and radiological issues, air pollution meteorology, and national and international cooperative programs. During the International Geophysical Year of 1957–58, he directed the U.S. research program in Antarctica and articulated the moral obligation of geoscientists to monitor the health of planetary systems. He also issued clear and strong warnings against uninformed climate engineering. Wexler served in uniform during World War II and consulted at the highest levels of national security during

Figure 4.1
Harry Wexler and Carl-Gustaf Rossby at Woods Hole, Massachusetts, May 1956.
Source: Wexler Family Papers.

the Cold War. He was the U.S. chief negotiator in Geneva for the World Weather Watch. Like his mentor, Rossby, Wexler worked at the cutting edge of geophysical research. He introduced a number of transformative technologies into meteorological practice that helped cut into, if not through, the Gordian Knot of prevision. After 1957 he nurtured the emergence of atmospheric science.

Young Wexler

Harry Wexler was born in 1911 in Fall River, Massachusetts, the third of four sons of Russian emigrants Samuel and Mamie (Hornstein) Wexler. His paternal grandparents, Isaac and Esther, had come to the United States in 1891 from Bălți, Bessarabia, in the Russian empire.[2] The name Wexler (originally Wechsler) referred to a person who changes things in the widest sense and later became associated with changing money (*geld wechsler*). True to his name, Harry Wexler changed things.

Harry's older brothers Charles and Max ("Mac") were talented in science as was his younger brother, Raymond, who became a meteorologist. All four attended Durfee High School, and all four excelled in academics. Harry's extracurricular activities included the yearbook staff and the chemistry club. He enjoyed "mathematical recreations" with his brothers and acquired a passion for science through his physics teacher, Leslie W. Orcutt, who was also his baseball coach. Wexler said his fascination for meteorology developed while delivering newspapers, intact and on time, through fair weather and foul. He shared an interest in weather with his childhood friend and neighbor, his future brother-in-law and colleague, Jerome Namias.[3]

Wexler enrolled in the SM program in meteorology at the Massachusetts Institute of Technology in the department of aeronautical engineering under the mentorship of Rossby, with coursework from Hurd C. Willett and Bernhard Haurwitz. He met Jacob Bjerknes when he visited MIT and attended his lecture series.[4] Experimental aircraft soundings and the new technology of balloon-borne radiosondes to measure vertical profiles of atmospheric temperature, pressure, and humidity were under development at MIT and undoubtedly made an impression on Wexler. He learned Bergen school methods of air mass and frontal analysis and the new techniques of isentropic analysis. He worked as a research assistant to Charles Franklin Brooks at the Blue Hill Observatory, just south of Boston. With data collected at this facility, Wexler published articles on the turbidity of North American air masses during the dust bowl years and recorded seasonal variations and examples of air masses changing their transparency as they traveled.[5] He began his lifelong affiliation with the Weather Bureau in 1934 as an assistant meteorologist in Chicago.

I Love You, Harry

Intimate details in scientific biography are often either missing from the record, unavailable, or scrubbed in the editorial process. The Wexler family papers contain Harry's letter of proposal to his future wife, Hannah Paipert, sent in November 1934 from his temporary duty station in Chicago (figure 4.2):

Figure 4.2
Harry and Hannah Wexler in Chicago, 1934. *Source:* Wexler Family Papers.

Sat. eve. 5:45
Nov. 24, '34
Dearest Hannah,
I just got home and found your letter. After reading it I did a very strange thing. I threw myself on my bed and wept as I did when I was a kid. The explanation for this peculiar performance is as follows:

I like Chicago quite a lot. My work too is coming along famously. I analyze my map, have a map discussion with Ray, and get home about 2 or 2:30 p.m., (I go to work about 7:15 or 7:30 a.m.). Then to kill time, I go to the University of

Chicago Library nearby to read some stuff on meteorology. Then afterwards to prevent myself from "thinking," I buy the fattest newspaper I can get and read it while dining at the very nice restaurant across the street. Then I go to my room, which by the way is very comfortable—and read for a while, and then to bed. Unfortunately, here, I can't escape my thoughts by reading, and so I lie sleepless for many hours—tossing around—thinking of you and wanting you here with me. I've fought against it—realizing that I'll be with you the middle of January—but I can't help it—I want you here with me now. The young fellows at the office are damn considerate. They realize I'm a stranger and have invited me out to meet some girls. I refused. This afternoon I went to a movie to see the Count of Monte Cristo— I still don't know what took place on the screen. I have no desire whatever to take walks—alone—without Hannah. In short, no matter how much I've tried to keep it curbed, I must ask you whether you can leave to come to Chicago at once and be married to me here. I know there are several reasons against this idea—it has taken me a week away from you to realize that they are all trivial beside the fact that I want you here with me at once. The objections to this proposal and a discussion follow.

1. *Money*—I have some $50. On Dec. 8, I will receive $96 and again on Dec. 22. The room I have costs $4 weekly—it is large, has a double bed and a private washbowl,—but no private bath. Tomorrow, I am going to look for a room (or a furnished apartment) costing not too much with a private bath.
2. *Your mother's* objections as to my haste and why can't I wait, etc., etc.—to this I can only say that I love you more than anything in the world—even more than my work, and I need you here now.
3. *What you can do here*—This section is the art section. For example, there is an art school nearby. This room has been used by an artist. Maybe in our new room (if I can find one reasonably enough) you can also work in the mornings. In the middle afternoon, I shall return and then we can go for walks, visits to art museums, etc. together.

Now, Hannah Darling, please consider the question carefully. If you see any flaws in your coming here immediately as my wife let me know, so that maybe I can figure their solutions. You don't have to bring much stuff so don't take much time for packing. You can have the other stuff sent to you. Then also when we go to Washington in January, we may be able to visit the folks and arrange to have the permanent stuff sent to us in Washington. If you decide to come, let me know immediately. You can make it comfortably by sleeper. Try to get a train that gets here in the afternoon or evening so that I can meet you. Or anytime Sunday. So please let me know Hannah dearest and please try to come to me.

I love you, Harry.[6]

The couple had two daughters.

Air Mass Analysis

In 1935, the new Weather Bureau chief, Willis Ray Gregg, established an Air Mass Analysis Section in Washington, DC, under the leadership of Horace Byers and assigned Wexler to help with the task of developing operational techniques of air mass and frontal analysis. His duties involved leading map discussions and teaching the new methods to others, including older forecasters who were already set in their ways. By this time, it was quite clear that, compared to the North Sea and Norway, where the Bergen school methods were first developed, air mass analysis was much more complex over the North American continent. Beyond the task of standardizing and objectifying the techniques of weather map analysis and forecasting lay a greater challenge in understanding the basic dynamics of atmospheric flows and their interaction on continental, hemispheric, and global scales.

Wexler's career parallels the research agenda for global meteorology in the twentieth century. He advanced the collection of comparable, simultaneous measurements and studied the interactions between high and low latitudes and between the Northern and Southern Hemispheres. Evidence of such work appears in his 1935 article on "Deflections Produced in a Tropical Current by Its Flow over a Polar Wedge" in which he examined a theory of Vilhelm Bjerknes linking events in the stratosphere and lower troposphere. He pointed out that Bjerknes's model did not agree with observations and did not explain how the north-south deflections in a tropical current could have amplitudes as great as 1,600 kilometers. Wexler suggested some rather simple adjustments to the theory. For Wexler, this was but the first in a long series of studies of the general circulation of the atmosphere.[7]

ScD

In 1937, the U.S. Department of Agriculture's Bankhead-Jones Special Research Fund provided funding for Wexler to return to MIT for further study. In Cambridge, he continued his research on polar air masses, supported Rossby's program on the general circulation, and developed techniques useful in extended or long-range forecasting, an interest he shared with Namias. As evidence of the inroads Bergen school techniques were making at the Weather Bureau at the time, Wexler had to pass a series of

"Air Mass Analysis" examinations to be promoted to the rank of associate meteorologist. In the oral defense of his thesis, he analyzed surface maps, upper-air data, and two vertical cross-sections (west-east and north-south) of a warm front. He also explained the principle by which cooling of the surface and the atmosphere extends to higher levels, generating a polar air mass. Here Wexler displayed complete mastery of the theory of atmospheric infrared emission by carbon dioxide and water vapor as applied to the evolution of air masses, especially how cold they can actually get.[8] Anticyclones became his specialty, and he followed this study with an analysis of aerological observations of outgoing radiation at high latitudes as part of his investigation of the formation and structure of polar continental air.[9] In 1938, Wexler was promoted to associate meteorologist in the Weather Bureau and in February 1939 passed his oral examination for the MIT doctorate.

Two weeks later, Hannah and Harry Wexler departed on a month-long cross-country auto trip to Arizona and California. They stopped at the Grand Canyon and Saguaro National Monument in Arizona, visited the movie set of *Gone with the Wind*, and explored Yosemite National Park. Harry's assignment was to discuss atmospheric radiation and possible Bankhead-Jones funding with Arthur Adel at the Lowell Observatory and Walter Elsaesser at Caltech, and his aim was to share information and develop closer working relations across disciplines. These two men were the leading experts on the spectra of water vapor, carbon dioxide, and ozone and the fundamental roles that absorption, emission, and transmission of radiation play in meteorological processes.[10] Adel had received his doctorate in physics from the University of Michigan, where he had studied the transmission of stellar and solar radiation through Earth's atmosphere. He was working at Lowell Observatory in 1939 to correct their observational data for atmospheric effects. Elsaesser, a German-born physicist and geophysicist, was using the MIT differential analyzer to construct a diagram useful for computing radiational cooling due to carbon dioxide and water vapor at various heights in the atmosphere.

Wexler presented two lectures at Caltech on the fluid mechanics of the atmosphere. He recounted how, at the turn of the twentieth century, Vilhelm Bjerknes and his associates had replaced the ideal fluids of Helmholtz and Kelvin with conditions closer to the actual behavior of the atmosphere but were having difficulty including friction in their equations. Anyone who has observed sky writers, for example, knows that there is mixing aloft, which implies the presence of frictional forces. The MIT group took a semiempirical approach that involved observing moving fluids in a wave tank and, by analogy, sought explanations of the flow patterns observed in the atmosphere as a combination of thermal, frictional, and topographical influences.[11]

In June, Wexler formally received the doctor of science degree from MIT; his dissertation, supervised by Rossby, combined observations of transverse circulations in the atmosphere with theoretical insights into their climatological implications.[12] He was reassigned to Washington, DC, to work with Rossby, who by then had become assistant chief of the Weather Bureau in charge of research and education.

Wexler's talents were much in demand. In the summer of 1940, he spent three months at LaGuardia Field serving as chief forecaster, tasked with introducing new methods of analysis, both at the surface and aloft, for New York and New England and for flights across the Atlantic. He performed similar tasks at other airway and district forecast stations throughout the country—in the San Francisco Bay Area in support of transpacific flights, and in Jacksonville, Florida, where he got a taste of tropical weather analysis and hurricane forecasting. In September, he was about to take up an assignment in New Orleans when Byers and Rossby offered him a job as assistant professor of meteorology at the newly established Institute of Meteorology at the University of Chicago. His duties there, while on a leave of absence from the Weather Bureau, included instruction of thirty-five advanced students in meteorology as part of a new national defense program and research on the general circulation with particular reference to five-day and longer-range forecasting.[13]

The Wexlers spent only one year in Chicago. They found the gray skies and brutal winters unappealing, and in late autumn 1941 Harry returned to the Weather Bureau in Washington, DC, as senior meteorologist in charge of training and research. His duties included staffing, recruiting, placing students in training programs, providing technical advice to the historical maps section, and serving as a liaison between the military and universities for research and educational purposes. Wexler provided consulting, text, and illustrations for an animated classroom film called "The Weather" produced by Encyclopedia Britannica Films.[14] It presented a study of the weather and techniques for measuring it and predicting its changes. The

Figure 4.3
Dynamic Meteorology Class 43–2-C, Army Air Forces Technical Training Command, Grand Rapids, Michigan, August 1943. *Source:* Wexler Family Papers.

film introduced the basic meteorological instruments, air mass and frontal analysis, and the general circulation of the atmosphere. A case study traced a wave cyclone on the polar front across the Midwestern states and related it to observations.

Wexler accepted a commission as a captain in the U.S. Army in 1942, serving as a "weather officer" under General Ben Holzman, director of the Air Force Research Center in Cambridge, Massachusetts, which was responsible for a program of weather research to meet the needs of the U.S. Army Air Forces. Wexler then served as the senior instructor of meteorology at the Aviation Cadet School in Grand Rapids, Michigan (figure 4.3). In this capacity, Rossby invited him to join the University Meteorology Committee, which was established to coordinate training and assist the military services (chapter 3).[15]

Following his promotion to major in 1943, Wexler moved to the new Pentagon building, where he worked in the weather division, U.S. Army Air Forces headquarters, in charge of research and development. The Joint Weather Central Office produced short- and long-range forecasts in support of the war effort. New technologies for research were being opened up by discoveries in electronics and by improvements in aviation and the extension of flight to new altitudes and into new areas of the world. The task at hand was to utilize meteorology more effectively in aerial navigation, bombing ballistics, and weather forecasts for military operations.

Wexler participated in a reconnaissance flight into a hurricane in 1944.[16] As Wexler recalled in 1962,

The flight took place out of an innocent remark of mine on the morning of September 14th, 1944 when I was aware of the hurricane off the coast and suggested to Col. F. B. Wood, with whom I was associated at the Headquarters, AAF Weather Division at the Pentagon, that since he usually takes Thursday afternoons off for his flying time, why not fly into the hurricane? Whereupon I proceeded to give a good hearty laugh and promptly forgot about it. But right after lunch, Col. Wood showed up and asked if I was ready, and I said "Ready for what?" and he said "Why, to fly through the hurricane." I said, "Are you kidding?" He said, "No, isn't that what you suggested?" I had to admit that I did. He thought it would be good also to have a forecaster-type come along. He [recruited] Lt. Frank Record to come along. . . .

At about 2:00 p.m., they flew in a Douglas A-20 Havoc from Bolling Field in a southeasterly direction for about 240 kilometers at an altitude of 900 meters, beneath an overcast sky and with bands of precipitation below them: "We could see the turbulent seas below with long white streaks of foam with larger tankers being washed from bow to stern by the ocean." Then they turned northeast, entered the storm, and flew another 160 kilometers into it through intense precipitation. After reaching the edge of the eye wall, a clearer space at the center of the storm, the pilot reported the plane suddenly gained 600 meters in altitude, and he had to nose it down to avoid being carried even higher. They were back at the airfield by 5:00 p.m. after an exhilarating flight. Wexler said nothing about the adventure when he arrived home, but Hannah noticed he looked a little pale and needed a drink. His efforts to keep it secret from her came to no avail when the story of the flight broke in the newspaper. At the Weather Bureau, he came to be known as "Hurricane Harry." Wexler protested that, without any instruments along, this probably was "the most un-scientific flight ever made into a meteorological phenomenon." The newspapers insisted on reporting this as the "first" flight into a hurricane, but it was widely known that Colonel Joseph Duckworth had flown into a hurricane off the Texas coast the summer before. Finally, Colonel Wood in exasperation issued a disclaimer: "Well, all we do is claim that this is the only flight made by three desk-bound officers leaving the Pentagon Thursday afternoon, September 14th and going into the storm and coming back!"[17]

By war's end, Wexler's portfolio included a number of new and potentially significant technologies relevant to atmospheric research. During the final four months of 1945, he worked to tie up loose ends and ensure continuity in the research programs. Wexler received one more promotion, to the rank of lieutenant colonel, before leaving the army and returning to the Weather Bureau in January 1946 as chief of the scientific services division.[18]

In the Catbird Seat

"Sitting in the catbird seat" refers to a superior or advantageous position or perspective, which Wexler surely attained in 1946 when he became director of meteorological research in the U.S. Weather Bureau. His responsibilities included planning and directing the overall theoretical and experimental research program of the bureau, including the development of new concepts, methods, data, and procedures in meteorology and related fields. His duties involved oversight of the research budget and personnel, serving as a liaison to government, universities, foreign governments, and international organizations, and participating in special national and international programs.[19]

As head of research, Wexler encouraged the development of new technologies, including nuclear tracers, weather radar, sounding rockets, digital computing, weather satellites, and space probes. He served on high-level committees in all these areas. He also maintained an active research program on breaking and pressing issues, such as the effects of nuclear detonations on the atmosphere, environmental concerns regarding nuclear reactors, air pollution meteorology, and weather and climate modification. He was in constant circulation as an adviser and collaborator in academic, governmental, and international circles. He was an effective communicator, widely sought as a spokesperson for geophysical research. His voice was heard from the dais at countless special events, internationally on Voice of America, and on the NBC television network.[20] He was a prominent and influential public servant who nurtured the multiple high-tech revolutions that shaped modern atmospheric research.

The Second Atomic Age

The second atomic age, inaugurated by the nuclear detonations of 1945 and subsequent atmospheric tests, raised new issues and generated new areas of responsibility for meteorologists. How would atomic bombs affect the weather? Where was the fallout transported, and how did radioactivity influence human health? Wexler and his associates took the lead in studying these issues. He developed techniques for following radioactive tracers downwind and around the globe, even in the snowfields of Antarctica, and studied the weather's effects on reactor safety. He established the "Special Projects Unit," a euphemism for research on classified nuclear issues; supervised the Weather Bureau publication *Meteorology and Atomic Energy*; and provided expert guidance to Congress and the public.[21]

While serving in the military, Wexler and L. W. Sheridan analyzed the pressure wave generated by the Trinity atomic test of July 16, 1945, as recorded by barographs located up to 400 kilometers away. The data indicated the disturbance did not travel along the earth's surface but instead skipped some areas and traveled outward in an indirect route, bouncing off of higher layers in the atmosphere.[22] At Hiroshima, Japanese meteorologist Michitako Uda observed meteorological conditions during the atomic bombing of his city on August 6, 1945, and provided an eyewitness account of the firestorm and black rain.[23] Weather played a much different role three days later when Kokura, the primary target for the second atomic bomb, was spared due to the industrial haze and smoke obscuring the target. Instead, Nagasaki, the secondary target, was obliterated when a brief opening appeared in the overcast. No firestorm occurred there, nor was any black rain reported.

The public was fascinated by the power of atomic bombs and wanted to know if they could be used to dissipate hurricanes and tornadoes or otherwise improve the weather. At the time, Irving Langmuir was employing nuclear metaphors to promote "weather control," for example, he claimed that "nucleating" agents could trigger "chain reactions" in clouds that could release as much energy as an atomic bomb. After 1951, when testing began in Nevada and mushroom clouds started to form over American soil, letters began to pour into the Weather Bureau and the Atomic Energy Commission blaming the unsettled, unusual, or unpleasant weather on atomic explosions. A large sector of the public was saying "Stop the tests. You're changing the weather."[24] They blamed nuclear explosions for tornadoes and drought in the United States, floods in Germany, and climate warming in the Northern Hemisphere. A Weather Bureau study comparing pre- and post-1945 weather records and a poll of leading meteorologists served to discount direct connections between atomic testing and the weather.[25] In his 1955 testimony to a Joint Congressional Committee on Atomic Energy, Wexler alluded to the natural human tendency to associate two events occurring close together as cause and effect and the tendency to connect extreme weather events to portents in the sky or on Earth. He argued that the geophysical energies shaping weather were so huge that a nuclear detonation on Earth's surface would be like a burning match tossed onto the sand of the Sahara desert: "The flaming match seems big to a nearby ant, but it adds very little heat to the desert—and it does not make the Sahara any dryer or any wetter."[26]

Radioactive clouds generated by outdoor nuclear testing introduced new tools that cut across the vast spectrum of atmospheric processes and, as Wexler put it, provided meteorologists with a "matchless tracing agent" of horizontal and vertical flow and diffusion applicable at all scales and all levels, from the surface layer to the stratosphere.[27] Where was the radioactivity transported and in what concentrations, on large and small scales? These were technical problems that only meteorologists could address with any specificity. In testimony to Congress, Wexler pointed out, "As a result of the Nevada tests, sufficient easily-identifiable tracer material is introduced into the atmosphere to be followed across the country by sensitive instruments." Observations of this type had supported studies of the paths of air masses and the rates of atmospheric mixing. Rainout of radioactive debris also served as an indicator of precipitation processes, and the data were useful for predicting fallout patterns for civil defense purposes, "so that danger to the civilian population in time of nuclear war may be minimized by directing evacuation into areas not affected by wind transport of radioactivity."[28]

At the time, there were two distinct communities of meteorology, the micro and the macro, working at different scales and in different institutions. Scientists at the national atomic laboratories in Brookhaven, Oak Ridge, and Idaho Falls were mainly pursuing micro problems, becoming masters of all things local. They were reading works on the atmospheric boundary layer by Ludwig Prandtl, Rudolph Geiger, and Oliver Graham

Sutton, and they were collaborating with engineers for the proper design and siting of their sensors. The macro people—mainly in Cambridge, Chicago, Princeton, and Washington, DC—were working on trajectory analysis on a hemispheric basis. They studied air masses, jet streams, and upper-air westerly waves, reading primarily the papers of Carl-Gustaf Rossby, Erik Palmén, and Jule Charney.

Wexler's close associate, Lester Machta, bridged the micro and the macro. He served during the war as an instructor in the Grand Rapids meteorological training program. After receiving his ScD from MIT in 1948, he joined the U.S. Weather Bureau, where Wexler assigned him to the Special Projects Unit. One of his first tasks was to coordinate with the Air Force Office of Atomic Energy, AFOAT-1, in an attempt to detect Soviet nuclear tests and compute the trajectory of their debris cloud.[29] To do this, Air Weather Service B-29 aircraft equipped with special air filters flew repeated missions between Alaska and Japan—that is, downwind of the Soviet Union. The filters detected radioactive particles on September 3, 1949, just east of Kamchatka at the 500-millibar level. Machta's job was to calculate backward trajectories and, from analysis of the radioactive decay of the debris, estimate the time and location of the test. Soviet test RDS-1 (called Joe-1 by the United States) occurred on August 29, 1949, at the Semipalatinsk Test Site in Kazakh (now Kazakhstan). Using trajectory analysis, Machta and his team calculated a detonation date two days prior and a detonation location near the Caspian Sea, all within the error limits of the observations they received.[30] The Soviet test sent a shockwave through the Western world because it was not expected to occur as soon as it did. On the other hand, the rapid detection of the test by the West shocked the Soviets, who had hoped to keep it secret. Wexler and Machta had developed detection techniques based on earlier Pacific tests by the United States where the timing was known and the paths and concentrations of the radioactive clouds of debris could be calculated from weather maps. Such trajectory studies of radioactive debris helped meteorologists understand the atmosphere better and served as a stimulus to general circulation modeling (figure 4.4).

A pressing issue was whether the Soviet Union could inject fallout into rainclouds headed over U.S. territory. Wexler and Machta concluded they could not—at least not in any practical way—without detection. To deal a damaging blow via radioactive precipitation, an estimated explosive power of 10 megatons would be needed, delivered in one thousand separate

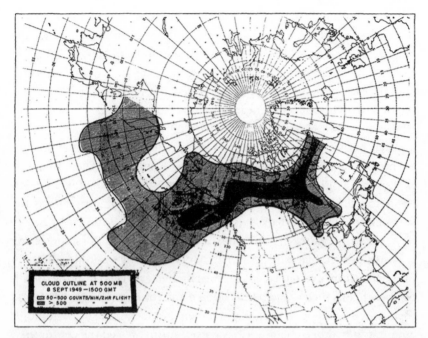

Figure 4.4
Spread of radioactive particles from the first Soviet bomb test as monitored by the U.S. Air Force and the U.S. Weather Bureau. From Machta, "Finding the Site of the First Soviet Nuclear Test."

10-kiloton detonations over the Pacific Ocean immediately upwind of the United States—an act tantamount to war that could not be disguised as a testing program.[31]

Machta spoke on "Nuclear Explosions and Meteorology" at a 1959 conference on Scientific Applications of Nuclear Explosions. The use of radioactive debris to track air masses was foremost on his list. He wrote, "We have been able to verify air trajectories computed from conventional winds, study lateral mixing as the clouds from Nevada atomic tests move across the United States, and watch the debris from the U.S. Pacific tests mix throughout the world."[32] Machta explored the possibility that meteorologists should call the shots so that bomb tests would be optimally located and properly timed, with identifying tracers added so that they could study particular features of planetary wind systems. With a nod to weather control, he speculated that the intensely heated bubbles of air generated in a nuclear explosion "can possibly be used to produce artificial water clouds" or that ionized debris from the tests might allow scientists "to change the atmospheric electrical properties at will." He thought detonating nuclear bombs in thunderstorms might help scientists understand both convection and electrical processes. He also supported the idea that such weapons might be used to redirect hurricanes on less damaging courses. Machta concluded that nuclear explosives could be profitably used for meteorological research but were too expensive and generated too much contamination for meteorologists to use them in their own experiments. There would be no nuclear Weather Bureau. Business as usual—that is, hitching a ride on the detonations conducted by the Atomic Energy Commission—was the best that could be expected.[33]

As powerful hydrogen bomb tests injected debris into the stratosphere, scientists learned that radioactive fallout circled the globe at high altitudes in the jet stream and rained out in concentrated events each spring. Insidious new toxins such as radioactive iodine and strontium-90 quickly found their way into the food supply and the environment. The detonations of atomic and hydrogen bombs in near space further fueled the controversy. The international outcry over atmospheric testing led to the First World Conference Against Atomic and Hydrogen Bombs in 1955, held in Hiroshima on the tenth anniversary of the bombing. In 1963, following a long crusade spearheaded by nongovernmental groups and supported by meteorologists and other scientists, the three nuclear nations—the United States, Great Britain, and the Soviet Union—signed and ratified the Limited Test Ban Treaty, which banned atmospheric nuclear testing in the atmosphere, underwater, and in outer space.[34] This marked the end of the second atomic age.

Air Pollution Meteorology

Let us use the atmosphere intelligently and not dump noxious material into it regardless of the diffusing efficiency of the atmosphere.
—Harry Wexler, 1950[35]

Wexler's first excursion into air pollution meteorology followed a devastating incident in October 1948 in Donora, Pennsylvania, "home of the world's largest nail mill," when industrial smog trapped in the deep

Monongahela river valley claimed the lives of twenty people and sickened thousands more. Wexler joined a study sponsored by the U.S. Public Health Service and served as a coeditor of its report, "Air Pollution in Donora, Pa." At the time, it was the "most exhaustive [study] ever made on a problem in air pollution." Wexler concluded that the U.S. Steel zinc plant in Donora was emitting significant amounts of toxic pollution from its smelters, waste heat boiler stacks, sintering operation, acid plant, ore roasters, blast and open-hearth furnaces, and blooming mill. Domestic heating systems and local steam locomotives in the town of Donora also contributed to the general atmospheric pollution of the river valley. Wexler used his knowledge of anticyclones to explain why the weather conditions necessary for the air to stagnate and smog to build up over a town like Donora were most likely to occur in the month of October. He then outlined five dynamic conditions for a "deep warm anticyclone to form and persist over a given area," producing a thermal inversion, greater air stability and stagnation, especially at night, lower wind speeds in the valley, and a buildup in concentration of contaminants. Although there had been earlier claims against U.S. Steel, the study portrayed the 1948 episode as a rare phenomenon but nevertheless one that sounded a clear alert about the health concerns associated with air pollution. It contributed to the passage of the Air Pollution Control Act of 1955, the nation's first piece of federal legislation on the issue.[36]

Wexler often spoke of both the complexity and simplicity of air pollution meteorology. The complexity arises from the need to study turbulent diffusion and chemical transformations in the lowest levels of the air as it travels over rough, moist, and heated surfaces. Simplicity refers to the as-yet very basic models available to meteorologists. To improve present methods and gain some local experience, Wexler suggested that researchers should focus on city microclimatology by attaching instruments to tall radio towers and driving instrumented vehicles around a city. He alluded to public awareness of pollution problems triggered by the fatalities in Donora and questions of how best to deal with them. He thought that since the sources of effluent were local, the focus of study and mitigation belonged at the local level. This was to be supplemented by actions at the county, state, and federal levels if the pollutants crossed political borders. As examples, he cited advice and assistance from the Weather Bureau given to the cities of Cleveland and Los Angeles and to the Trail Smelter arbitrations in British Columbia.[37]

Wexler studied the atmospheric whirls and eddies that disperse air pollutants and recommended proper siting of smoke stacks and proper timing of pollutant releases when the air was well mixed. In his final publication on air pollution meteorology, Wexler reviewed the role of airflow in the dispersion and diffusion of pollutants and in scrubbing pollutants from the atmosphere. He was aware of chemical reactions in the atmosphere and transformation of pollutants, but studies of these phenomena were just getting underway.[38] Well before the modern environmental era, Wexler warned that that atmosphere should not be treated as a dumping ground for unwanted air-borne materials.[39]

Weather Radar

"Through the use of radar for precipitation detection the science of meteorology has acquired an entirely new and unique method of weather observation": so begins a 1951 article on radar in the *Compendium of Meteorology*.[40] It had long been known that metallic objects reflected radio waves. In 1930, engineers reported that weather conditions, especially rain and fog, had a noticeable effect on shortwave radio propagation.[41] By sending out short pulses of energy and by receiving and timing their reflections, they were able to determine the distances and directions to targets. After 1934, a number of nations developed such experimental radio devices independently and in great secrecy, but the long wavelengths employed resulted in poor spatial and directional resolution. By 1939, the acronym RADAR (for radio detection and ranging) was applied to these technologies.[42]

Radar beams are scattered by hydrometeors. This meant that patterns of rain and snow, previously invisible, were beginning to show up as echoes on radar screens. On February 20, 1941, the British military tracked a rain shower off the English coast using 10-centimeter radar, a capability soon duplicated at the MIT Radiation Lab.[43] What was initially seen as a drawback for detecting and ranging enemy targets soon became a new research tool for meteorologists. As Henry Guerlac notes in his monumental *RADAR in World War II*, "by August 1942 the application of radar techniques to weather analysis and forecasting appeared to be a distinct possibility, and the military services and the weather bureau expressed strong interest."[44] The possibility that vertically pointing radar could assist aviators is documented in a letter, classified secret, from Arthur E. Bent at MIT to Francis W. Reichelderfer at the Weather Bureau:

A development in our program here has occurred . . . that may require . . . the assignment of a Bureau meteorologist to work with us. . . . We have been able to identify certain atmospheric phenomena such as thunderstorms and active convective clouds, as well as certain types of rain. These identifications have been at distances of a few miles to 160 miles or more. While the location of these areas at a distance is of great importance, it was also felt that the indication of stratifications above an observing station would be of great significance. We have consequently adapted our equipment for operation in a manner to give such results and can now report that we are able, under certain conditions, to identify layers or areas directly above a station.

Bent requested a Weather Bureau employee who could launch radiosondes over Boston at the same time radar beams were actively probing for layers of rain and icing over the airports.[45]

The refraction and attenuation of beams directed horizontally provided fundamental new information about the absorption properties of water vapor and about humidity conditions over the moist ocean boundary layer. This electronic probe—in effect, a remote psychrometer—responded much more quickly and accurately than traditional balloon-borne meteorographs and radiosondes, allowing scientists access to the fine structure of the atmosphere and thus its micrometeorology.[46] The military used this technology to detect ships, submarines, and low-flying aircraft in the haze. Contemporaneously, MIT Radiation Laboratory physicist Robert H. Dicke invented his eponymous radiometer using a microwave antenna that serves as a thermometer that measures the temperature of the source of the radiation. This was a critical step in atmospheric remote sensing, with many other applications in other sciences.[47] To train weather officers in radar, the Air Weather Service conducted an intensive, seven-month radar course for new weather officers, with the first group graduating in October 1944.

Meteorologists quickly recognized the differences in radar appearance of showers and widespread rain, the association of precipitation echoes with fronts, and the characteristic echoes of thunderstorms and tropical storms, but research by established scientists was limited due to security constraints. As the war was winding down, the results of classified programs could be made public and new programs could begin. Navy Commander R. H. Maynard published examples of radar scope photographs of fronts and storms

obtained at Lakehurst, New Jersey, and of a tropical cyclone in the Philip-
pine Sea, explaining that "radar offers to the pilot and forecaster an actual
'picture' of storms," which can be analyzed to determine the "length,
depth, height, intensity, speed and direction of movement of any storm in
the area covered."[48] By February 1944, the U.S. Army Air Forces had
equipped many of its airplanes with microwave radar that enabled flight
crews to navigate around storm areas.[49] In an editorial note in the *Journal of
Meteorology*, Wexler lauded the use of radar for storm detection and tracking
as "one of the most effective meteorological tools developed during the
war." He continued,

No one knows who was the first to peer into the radarscope and decide that certain
echoes were caused by reflection from precipitation occurring in storms. It is certain,
however, that as early as 1942, personnel, both civilian and military, in all parts of
the globe, in universities, in laboratories, and in combat units on the sea, in the air,
and on the ground appreciated the tactical advantage to be gained by using radar in
detecting storms as far away as 150 miles. Meteorologists quickly applied the results
to forecasting, and by the end of the war radar storm detection was a vigorous new
branch of meteorological science. Therefore, in view of the enormous importance of
radar [in meteorology], it is most appropriate to give credit to all the pioneers in this
field who contributed to the vast amount of fundamental and operational knowledge of radar and weather.[50]

Wexler used radar to determine the structure of a hurricane that passed over
Florida on September 14–15, 1945. He compared traditional observational
techniques with time-lapse images and tracking provided by the high-
power, long-range radar at the U.S. Army Air Forces Center in Orlando,
Florida. Wexler included radar screen images of squall lines and hurricane
bands along with new explanations of their dynamics. Waxing philosophical, as was his style, Wexler discussed the two traditional views of the
weather—the local or microscopic view and the synoptic or macroscopic
view. He called them "eyepieces." The first, limited to less than 30 kilometers (for example, from an airport control tower), provides detailed visual
observations of clouds, precipitation, and haze. The second is a constructed,
impressionistic overview of the various surface data collected locally, analyzed by a trained forecaster, and depicted on a weather map (for example,
in an airfield flight briefing center). Such analyses can be as extensive as a
continent or even the entire hemisphere. Its strength is the overall synoptic
view it provides; its weakness lies in a lack of intricate details. Aircraft
weather reconnaissance can fill part of this gap, although no airplane flies

high enough to provide a truly synoptic gaze. Instead, they follow particular trajectories and provide discrete weather reports. Wexler pointed out
that the radar screen was now providing a new "eyepiece" thorough which
to view severe weather, at once both extensive and detailed. A new research
field involving radar in meteorology had begun.[51]

Soon, radar meteorologists were able to locate all precipitation instanta-
neously over several thousand square kilometers horizontally and at all altitudes from a single observing station, estimate its intensity and determine
the direction and speed of its movement, identify areas of turbulence and
wind shear, and chart the distribution of fall velocities at any level. Related
capabilities included estimating the heights of clouds and the freezing level,
tracking the paths of thunderstorms and hurricanes, retrieving upper-level
wind data even when overcast, and estimating water-vapor content and
temperature distribution in the vertical. An example of this was the "Thunderstorm Project," directed by Horace Byers at the University of Chicago,
which began studying storm patterns using radar units attached to research
airplanes even before the war officially ended.[52]

Harry Wexler's brother, Raymond, also was involved in radar meteorology through the U.S. Army Signal Corps. After a family dinner, Harry
thanked Ray for his discussion of the radar pictures he took of a cold
front in 1945. He mentioned that his colleagues in the Weather Bureau
analysis center were extremely interested in how radar was able to reveal
fine structural details of weather phenomena and the possibility that a
cold front was composed of a series of discontinuous overlapping line
squalls.[53] Ray remained active in this field, later evaluating the available
radars for their suitability and recommending appropriate units for widespread meteorological use. The Weather Bureau adopted a navy design, the
Weather Surveillance Radar (WSR-1) and its successors, and deployed the
first unit at Washington's National Airport in 1947 (figure 4.5). Over a
decade later, the WSR-57 became the operational standard for the U.S.
National Weather Service.

The heady successes of weather radar led to the "Theme Song of the
Sixth Weather Radar Conference," performed for the first (and perhaps last)
time in Cambridge, Massachusetts, in 1957:[54]

More data, more data,
Right now and not later.
Our storms are distressing,

Figure 4.5
Weather Radar Antenna. Original caption: "Radar equipment reveals rainfall hidden by clouds and is useful in detecting severe squalls and tornadoes at distances up to 50–100 miles." *Source:* NARA, Reichelderfer Records, box 5.

Our problems are pressing.
We can brook no delay
For theorists to play.
  Let us repair
  To the principle sublime:
    Measure everything, everywhere,
      All the time.
For data are solid,
Though dull and though stolid.
Consider their aptness,

Their matter-of-factness.
Theory is confusion,
A snare and delusion,
  A dastardly dare,
  A culpable crime.
  Measure everything, everywhere,
    All the time.
No need to be weary
Of the mysteries of theory.
We have only to look
At the data we took,
Immediately inspired,
Grasp the answers required.
  What are so rare
  As reason and rhyme.
  Measure everything, everywhere,
    All the time.
More data, more data,
From pole to equator.
We'll gain our salvation
Through mass mensuration.
Thence flows our might,
Our sweetness, our light,
  Our spirits full fair,
  Our souls sublime:
  Measuring everything, everywhere,
    All the time.

L'Envoi (Conclusion)
And it shall come to pass, even in our days,
That ignorance shall vanish and doubt disappear.
Then shall men survey with tranquil gaze
The ordered elements shorn of all fear.
Thus to omniscience shall we climb,
Measuring everything, everywhere, all the time.

Arrays of multiple radar units with overlapping coverage opened the possibility of a new, extended aerological network to improve aviation safety and provide detailed depictions of current weather in support of very short-term forecasts. Today this is called "nowcasting," Such enthusiasm for radar supports Rossby's point about the role of technology in "cutting the Gordian knot instead of untying it."[55]

Computers

Wexler learned of the possibilities of digital computing in 1945 and participated in a number of high-level conferences the following year. Serving as
Weather Bureau liaison and working with Rossby, he recruited meteorologists and mathematicians for the Institute for Advanced Study computer
project in Princeton, New Jersey. Later, Wexler took steps to institutionalize
and operationalize numerical weather prediction and general circulation
modeling.

In January 1945, at the birth of modern digital computing, University of
Pennsylvania physicist and engineer John Mauchly visited Washington,
DC, to explain to various research groups "the general nature and purpose
of the machines under development by the Moore School," including
the new EDVAC—the Electronic Discrete Variable Automatic Calculator.
Mauchly was seeking examples of challenging applications that would
guide the development of machines of maximum usefulness and flexibility.
At the Weather Bureau, he met with Charles F. Sarle, who focused on three
things—machine sorting of observational data on punched cards for later
climatological analysis, "extrapolating the weather map," and speeding up
statistical correlations. Mauchly remarked, "Although I mentioned several
times that EDVAC would be capable of solving partial differential equations, the use of the EDVAC in handling such equations when they arise in
meteorology was not touched upon until I asked about such possibilities."
Sarle offered one good piece of advice, though. He directed Mauchly to get
in touch with Major Harry Wexler at the Pentagon. The result was completely different. According to Mauchly, Wexler "displayed a great deal of
enthusiasm concerning the possibilities of the EDVAC in meteorological
research" and suggested immediately that such a machine be employed in
"integrating hydrodynamic equations occurring in meteorological work."[56]

Several months after Mauchly's visit, Wexler received a pamphlet on
"Modern Computing Devices" from RCA Associate Research Director Vladi-
mir K. Zworykin, whose most notable inventions included television transmitting and receiving devices. In September 1945, Weather Bureau chief
Reichelderfer visited the RCA Labs in Princeton, New Jersey, with other representatives of the Department of Commerce for a preliminary discussion
about the use of modern electronic devices in meteorological analysis. A
month later, Zworykin wrote his "Outline of Weather Proposal," at the time

influential but now all but forgotten. The report began by discussing the importance to meteorology of accurate prediction, which Zworykin thought was entering a new era. Modern communication systems were beginning to allow the systematic compilation of scattered and remote observations, and new computing equipment was becoming available that could either solve the equations of atmospheric motion or at least search quickly for statistical regularities and past analogue weather conditions. He imagined "an automatic plotting board" linking the computer to a television screen that would display all this information.[57] Zworykin suggested that "exact scientific weather knowledge" might allow for effective weather control. If a perfectly accurate machine could be developed that could predict the future state of the atmosphere and identify the precise time and location of leverage points or locations sensitive to rapid storm development, then intervention might be possible. A paramilitary rapid deployment force might be sent out to intervene in the weather as it happens—literally to pour oil on troubled ocean waters or use physical barriers, giant flame throwers, or even atomic bombs to disrupt storms before they formed, deflect them from populated areas, and otherwise attempt to control the weather. John von Neumann, who had been involved in the development of EDVAC and its predecessor ENIAC (Electronic Numerical Integrator And Computer) formally endorsed Zworykin's proposal.[58]

Reichelderfer extended an invitation to Zworykin, von Neumann, and others to gather in Washington, DC, on January 9, 1946, with Edward U. Condon, director of the National Bureau of Standards, and representatives of the army, navy, and Weather Bureau. Wexler was at this meeting, not to endorse any weather control ideas but to see how computing and meteorology might work together. The meeting introduced meteorologists to digital computing and nonmeteorologists to current methods of weather analysis and forecasting.[59] The following week, a second conference brought Mauchly and his colleague J. Presper Eckert to the Weather Bureau. A flurry of communications followed concerning plans for a project located in Princeton to build a computer especially designed for the needs of meteorology.[60]

In April 1946, Rossby sketched a complete outline of the electronic computing project and sent copies of it to von Neumann and Reichelderfer. He thought that ENIAC, at that time located at the University of Pennsylvania, "could be used to solve numerically the problem of the general circulation of the atmosphere." He provided budgetary guidelines, suggestions for personnel, strategies to involve the U.S. Navy Office of Research and Invention (ORI), and identified the need for a conference on theoretical issues. He nominated Wexler to lead the project as a "thoroughly competent senior investigator . . . with a thorough synoptic knowledge of such thermodynamic factors as ice, clouds and radiation."[61]

In May, the Institute for Advanced Study submitted a proposal to ORI for "an investigation of the theory of dynamic meteorology in order to make it accessible to high speed, electronic, digital, automatic computing." Not surprisingly, the proposal closely followed Rossby's outline. Wexler consulted with von Neumann on the availability of personnel and their salary levels, offered Weather Bureau library services in support of the project, and followed this up with numerous visits and phone calls.[62] He promised von Neumann at least a half-time commitment to the project and considered relocating to Princeton. On July 19, Wexler learned from von Neumann's assistant, Herman Goldstine, that the Institute had received a contract for the meteorology project and was making arrangements to acquire war surplus housing. Wexler immediately called a meeting in Princeton in August to discuss both the objectives of the project and practical arrangements, asking all involved to "come with some definite ideas regarding objectives, problems, and working procedures . . . to examine the foundations of meteorology, to solve the basic problems of the general circulation, and to improve our understanding of atmospheric processes."[63] The meeting, hosted by von Neumann at the Princeton Inn, August 29–31, was dominated by the MIT school, featuring talks by Rossby, Bernhard Haurwitz, Hurd C. Willett, and Jerome Namias. Also of note were two visiting scholars from Chicago invited by Rossby, one famous and one relatively obscure— Jule Charney and Jeou-jang Jaw. Charney eventually directed the Institute for Advanced Study's computer project and produced a successful machine forecast. His primary focus was to eliminate all hydrodynamic equations that were not meteorologically important and to find a numerical solution for those remaining. Jaw, from the Institute of Meteorology of the Academia Sinica, worked on atmospheric waves, instability criteria, and perturbation equations in a baroclinic atmosphere.[64] Although primarily engaged with modeling the dynamics of upper-air flows, attendees at the conference also discussed projects involving oceanography, polar circulation, radiation balance, and new field experiments to provide inputs into the models.

Wexler suggested measuring divergence in the atmosphere by tracking constant-level balloons by radio.[65] In December Wexler and Rossby convened a second significant conference in Chicago on research problems in meteorology.[66] Wexler presented "a list of lines of approach in meteorological research that should be pursued," including measurements of outgoing long-wave radiation at various heights, numerical forecasting of a smarter variety than the "brutal assault" method, and a concerted attempt to model the general circulation, with improved observational evidence to define initial conditions. Rossby added to the list the need to pay greater attention to interactions between the hemispheres.[67]

The Institute meteorology project got off to a slow start, with von Neumann ruing the lack of a stable research group. Wexler suggested adding Philip D. Thompson from UCLA, a "young, enthusiastic, intelligent, and hard-working Army Air Force weather officer."[68] He also recommended regular meetings with a large group of interested scientists because the problems they faced were likely outside the range of expertise of any small, handpicked team that could be assembled on site. Thompson joined the Institute project in 1946, followed by Charney and Arnt Eliassen two years later. Still, the staff in Princeton had yet to reach a critical mass.[69] In 1949, von Neumann tried to recruit Rossby to head the project, suggesting that he might become a permanent member of the Institute with all the privileges pertaining thereto. He expected that the Institute's electronic computing machine would likely be in running order early in 1950 and available for use on problems in theoretical meteorology. He asked if Rossby was available to join the Institute for a period of two years, "if suitable financial arrangements can be made. If such an arrangement meets with your approval, I am also empowered to make this a formal offer." Charney supported this effort, encouraging Rossby to take leadership of the meteorology project: "Needless to say, I sincerely hope that you will decide to come. You know as well as I that meteorologists will continue to be frustrated at every turn as long as they lack the mathematical ability to carry their physical arguments to their logical conclusion. I would like nothing better than to be able to help you break this dam." Wexler weighed in, too; he was convinced that the next major advance in forecasting would come from the Princeton group.[70] Institute director Robert Oppenheimer sent Rossby a formal offer of appointment in July 1949 for the upcoming academic year. Receiving no reply, he reissued the offer in September to cover the winter

Figure 4.6
Photo taken in front of the ENIAC, Aberdeen Proving Ground, April 4, 1950, on the occasion of the first numerical weather computations carried out with the aid of a high-speed automatic computer. Left to right: Harry Wexler, John von Neumann, M. H. Frankel, Jerome Namias, John C. Freeman, Ragnar Fjørtoft, Francis W. Reichelderfer, and Jule Charney. *Source:* Wexler Papers 23.

months. Rossby finally responded in October, asking that, if possible, the offer be extended for an additional year. He felt that the Princeton project was already in able hands and that he had more foundational work to do in Stockholm and Chicago.[71]

The Princeton team ran successful tests on the army's ENIAC in 1950 and 1951 at Aberdeen Proving Grounds, Maryland (figure 4.6). Although computational meteorology was still in its experimental phase, they were able to produce twenty-four-hour forecasts from actual meteorological data. Right after the experiment, Charney wrote to George Platzman at Chicago, "I think we have enough evidence now to bear out most of Rossby's prophecy."[72]

ENIAC was designed as a general-purpose machine to solve large systems of equations. Although it required laborious and time-consuming manual programming, once running, it outperformed the others, being a hundred times faster than MIT's differential analyzer and a thousand times faster than Harvard's Mark I. It could calculate in twenty seconds the trajectory of an artillery shell that took thirty seconds to reach its target. It was used to calculate hydrogen bomb yields. The machine was huge, arranged in the shape of a U some 80 feet long. Its circuits consisted of 18,000 vacuum tubes, 70,000 resistors, and 10,000 capacitors operating as on-off switches, or logical gates, with each digit on ENIAC represented by ten vacuum tubes.[73]

In April 1952, Rossby sent Charney a long account of the Swedish BESK computer and asked if it would be possible "to obtain from your group information or copy of the code for the barotropic model, as coded for the Princeton machine? It would save us a lot of work." He also sought the loan of recent Chicago PhD Norman Phillips for six months to help with programming.[74] In December, BESK produced experimental retrospective forecasts, and on March 23–24, 1953, BESK generated the first real-time numerical weather prediction, beating the actual weather by some ninety minutes. This involved replacing the time and space derivatives in the equations of motion for the upper air with differences of finite magnitude (for example, one hour for the time increment and 300 kilometers for the horizontal spatial grid). The meteorological problem consisted of programming the digital computer to compute the changes in the state of the atmosphere, hour by hour, from the values of the fundamental parameters. With a computer fast enough to complete these calculations in less than one hour, the expected changes could be added to the originally observed values to generate, via an iterative or "bootstrapping method," a twelve-hour or longer prognosis. This was the method used in routine numerical weather prediction services. Rossby, traveling in the United States at the time, received a telegram from Stockholm: "First operational NWP carried out last night." BESK also produced operational forecasts in the autumn of 1954 during Operation Dalamanövern, a military exercise in central Sweden emulating the results of a nuclear exchange, a situation in which the upper-air wind trajectories were of utmost importance.[75]

Charney had been at Princeton since 1948, but there was no clear path for his promotion. Von Neumann proposed permanent membership for him in the Institute. Jacob Bjerknes wrote a surprisingly lukewarm letter to von Neumann in 1951. He had a "favorable" impression of Charney and expected him to make valuable contributions "in the time to come," but opined that "the real proof for Charney's greatness is still lacking although we expect it to be forthcoming." He ranked him third after Norwegians Arnt Eliassen and Ragnar Fjørtoft, whose scientific records were "well ahead of Charney's." He said UCLA's policy would be not to promote Charney at this time until more proof is available, but he then waffled, saying that a promotion for Charney in Princeton would be "very timely."[76] When the Institute's new electronic computer, called the "Johniac" after John von Neumann, became available in 1953, Charney led the effort to provide regular forecasts and verifiable hindcasts. Late that year, using a three-level model running on the Princeton machine, Charney successfully modeled a previously unforecast storm that had dropped six inches of snow on Washington, DC. He immediately called Wexler, awaking him from a sound sleep, and informed him "Harry it's snowing like hell in Washington on November 6, 1952."[77]

As the initial experimental phase of numerical weather prediction was winding down, Wexler represented the Weather Bureau in meetings convened by von Neumann aimed at making it operational.[78] Recent modeling successes encouraged the formation in the United States of a Joint Numerical Weather Prediction Unit under the coordination of Wexler at the Weather Bureau with the participation of the air force and navy weather services and with a number of consultants including Charney, Rossby, and von Neumann. George Cressman directed the new effort with Philip D. Thompson in charge of the development section. A rented IBM 701 computer arrived in Suitland, Maryland, early in 1955, and after testing and trial runs, the facility issued an operational numerical forecast for thirty-six hours on April 18, 1955. U.S. press coverage of the story ignored the earlier success of the Swedish BESK computer and emphasized that it was the first time that weather forecasts would be made from physical equations reflecting atmospheric conditions, thus fulfilling the half-century-old vision of Vilhelm Bjerknes.[79]

All was not well in Princeton, however. Von Neumann had trouble getting the Institute to accept meteorology as a department and ultimately failed in this effort. In 1954, mathematician and physicist Freeman Dyson collected outside opinions regarding the proper role for the Institute in the fields of applied mathematics and electronic computing. The "computer project" was meant to be the nucleus for a group of applied mathematicians to use the machine for research into the mathematical theory of complex nonlinear systems of equations. It was not quintessentially or only a "meteorological project." Dyson proposed three alternatives—(1) incorporating the existing meteorology group into the permanent Institute organization, (2) extricating the Institute from direct support of the computer and letting the government run it as a practical project, and (3) letting the meteorologists go elsewhere and keeping the computer for the Institute in the hope of establishing a school of fundamental research in some other branch of applied mathematics. Option two prevailed, at the government-supported operational numerical weather prediction unit in Suitland, Maryland.

Computers were getting faster and more powerful, but results so far had been obtained only by simplified equations for upper-air motion. It was not yet possible to use the primitive equations of motion in numerical weather prediction. Suitland director George Cressman, with access to the best computers, remarked that "the machine makes fine forecasts of upper-air weather for high-flying aircraft. For ground-level weather, it is not yet very good."[80] The long-range forecasting problem would have to include the functioning of a baroclinic, diabatic atmosphere, with turbulence treated realistically. More than just a problem of numerical method, it would require a new class of models. In this work, faster machines would be useful, but research on conceptual models would dominate.[81]

The digital computer is a tool, a powerful calculating machine for investigating theoretical models and for solving complex problems involving large quantities of data and arithmetic that lie beyond the scope of hand computation. It came into its own as a heuristic tool and a means of assimilating and visualizing observations, just as Bjerknes's hand-drawn streamlines had done a half century before. In Sweden, the BESK computer was used to analyze, sort, and check the 500-millibar observations and project them on the screen of a cathode ray tube. In Washington, teleprinter tapes containing the observations were introduced directly into the electronic computer, which sorted the data, checked them for internal consistency, discarded faulty observations, and then stored the data in the machine for final analysis. In these ways, the practice of objective data analysis was being automated. Although still shy of being a practical reality, the "vast machine" was gearing up.

In the summer of 1955, von Neumann, Wexler, and Charney met for lunch to develop a "Proposal for a Project on the Dynamics of the General

Circulation." Wexler suggested the title, developed the budget, and was slated to direct the project with von Neumann and Charney as consultants. The proposal succinctly summarized successes in "theoretical and computational investigations in meteorology," from the 1946 origins of the Princeton project to the "very good success" in operational numerical weather predication attained recently. Norman Phillips had just used the Institute computer to prepare a numerical simulation of the general circulation for thirty days, producing suggestive but incomplete results. With a model containing nonlinear dynamics, static boundary conditions, and simple energy sources and sinks, Phillips was able to produce realistic patterns—trade winds, westerly winds, moving cyclones and blocking highs, and above them, strong jet streams—before they broke down due to computational instability. If this was possible, what then was not possible?[82]

Short-term weather forecasts focus on the partial rotation of a single vortex. This is true both for Bergen school methods and for numerical weather prediction. The early models applied only to the large-scale phenomena of the middle atmosphere and excluded the effects of radiation, geography, topography, humidity, and even friction. Rossby said they employed "structure-dissolving" techniques that provided a "frog's-eye view" of weather.[83] Intermediate-range forecasts of thirty to ninety days, the holy grail of rational planners, need to work on larger scales—including all the diabatic, turbulent, and other nonlinear factors—to predict the future positions and intensities of multiple vortices, which was an intractable computational problem. More tractable was an attempt to integrate the atmospheric equations for a dynamic, moist, heated atmosphere *forever*—the so-called infinite forecast. The result was not actually a forecast or a prediction of particular weather conditions but a limiting case converging on the statistical features of the general circulation, independent of whatever initial conditions may have existed. Charney published an article on the "infinite forecast" in *Dynamics of Climate*, a book based on a 1955 planning conference held in Princeton. Modern climate models are distant (very distant) cousins of this process.[84]

The proposal for a General Circulation Research Section, dated August 1, 1955, was approved as a joint venture of the U.S. Weather Bureau, Air Force, and Navy. It was colocated with the operational numerical weather prediction unit in Suitland, with Joseph Smagorinsky in charge. Both groups used the IBM 701 computer but worked on different time scales and different levels of complexity.[85] Margaret Smagorinsky, who worked with her husband in Suitland, recalled Wexler's overall leadership: "Oh, Wexler, he was easy to talk to. He was the head of our unit, actually, and he used to bring his girls into the office. Hannah, his wife, was an artist, and she'd be doing something artistic or was showing her work."[86] Wexler knew he had been involved in historic events. In 1961, he issued a memo to his staff to preserve and compile all the documents they could find regarding the computing project.[87]

Sounding Rockets

The age of artificial Earth-orbiting satellites officially began on October 4, 1957, with the dramatic and historic launch of *Sputnik 1* by the Soviet Union, but the space age has much deeper roots.[88] Robert H. Goddard developed liquid-fueled rockets and used them for weather photography in the 1920s, radio telecommunications arrived in the 1930s, a V-2 rocket flight in 1947 photographed clouds from an altitude of 160 kilometers, and by 1954 instrumented sounding rockets had serendipitously photographed an unknown tropical storm (figure 4.7).[89]

Until the mid-1940s, upper-air measurements were taken by aircraft, with a ceiling of about 14 kilometers, and by sounding balloons that might reach 30 kilometers, After 1946, however, direct measurements of the upper atmosphere became possible due to developments in rocketry. Historian David DeVorkin examined how the V-2, Hitler's "vengeance weapon," was repurposed as the first vehicle to carry scientific instruments to extreme altitudes, verging on the edge of space. As a unique platform, the V-2 transformed civilian and military research agendas, fostered their convergence, and gave rise to new research specialties in space science as it shaped the landscape of Cold War American science. The V-2 was designed to lift a payload of one ton, much larger than needed for scientific experiments, to an altitude of about 120 kilometers. Properly instrumented, it could make direct measurements of temperature, density, and chemical composition and observe the solar spectrum above the ozone layer. Detonating a warhead at altitude produced a pressure wave whose propagation revealed information about density and temperature as well. About fifty captured V-2s were used for research; they were phased out by 1951, superseded by purpose-built sounding rockets.[90]

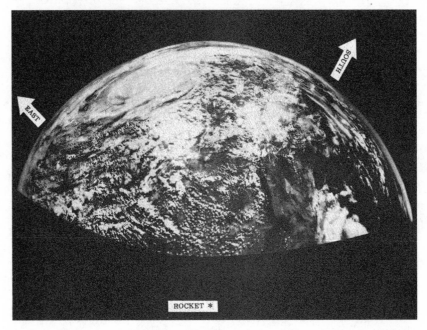

Figure 4.7
Image of a previously undetected tropical storm in the Gulf of Mexico photographed by an Aerobee sounding rocket in 1954. *Source:* Hubert and Berg, "Rocket Portrait of a Tropical Storm."

Viking, a U.S. Navy rocket developed for defense purposes, was about half the size of V-2. It could carry payloads as high as 240 kilometers, but it was expensive and unreliable. Only twelve ever flew. The Aerobee rocket was much less expensive, less complicated, and more reliable. It was used most frequently in the early days of upper-atmosphere research. Small, solid-fuel Deacon rockets were even cheaper, and when paired with a first-stage Skyhook balloon, these so-called Rockoons could reach an altitude of 115 kilometers with a 9-kilogram payload. Also coming on line in 1956 was a Nike-Ajax mobile antiaircraft missile, which, when paired with a small Cajun rocket as a second stage, could take vertical soundings up to 160 kilometers. The mobility of the launchers meant that small rockets could be fired from remote locations such as Fort Churchill, Canada, and from the decks of ships. Various rockets were used to study the pressure, temperature, density, and composition of the upper atmosphere; the solar spectrum in the extreme ultraviolet and x-ray regions; the ionosphere; Earth's magnetic field; auroral particles; and cosmic rays. One goal was to extend the standard atmosphere model from temperate latitudes into auroral and equatorial zones and conduct research on the geographical variation of other phenomena, from the Arctic to the Antarctic. Simple free-fall accelerometers were dropped at altitude to measure the density of the atmosphere, and more sophisticated electrical, magnetic, and radioactivity detectors flew during the International Geophysical Year.[91]

The rocket age introduced military considerations into upper-atmosphere research. In 1947, William Welch Kellogg proposed a thesis topic on the interdisciplinary science of the region. He had been an army air corps test pilot and meteorology instructor during World War II and now worked for the Rand Corporation while pursuing a PhD in meteorology at UCLA: "One is impressed, when reading through the literature on the upper atmosphere, by the diversity of the disciplines which have been brought to bear on this subject. It will require the combined efforts of scientists in nearly every branch of physics and geophysics to extract the knowledge that we now have. It is this very diversity which presents a challenge to anyone interested in adding to our knowledge of this region, and perhaps helps to explain why to date so little is known about it." Kellogg thought the increased interest in this region and its future importance were due to "the fact" that the upper atmosphere and near-space environment "would be a theater of operations in any future war, which certainly added a note of urgency to the normal leisurely course of scientific research."[92]

Rockets would soon be able to place satellites into Earth orbit. Kellogg and his colleague S. M. Greenfield opened their top secret RAND report of 1951, "Inquiry into the Feasibility of Weather Reconnaissance from a Satellite Vehicle," with the assumption that, "in the event of an armed conflict," an alternative method of obtaining weather reconnaissance over enemy territory, similar to that obtained in World War II, "is thought to lie in the use of the proposed satellite vehicle." In an appendix to that report, Jacob Bjerknes provided analysis of synoptic weather as observed from photographs taken on two rocket flights over White Sands, New Mexico, on July 26, 1948.[93] Sounding rockets came under Wexler's purview as scientific probes and observational platforms to investigate the upper atmosphere and to photograph clouds from above. Wexler served as chair of several influential committees on this subject, and his extensive records from meetings show the evolution of issues, including the official naming of the

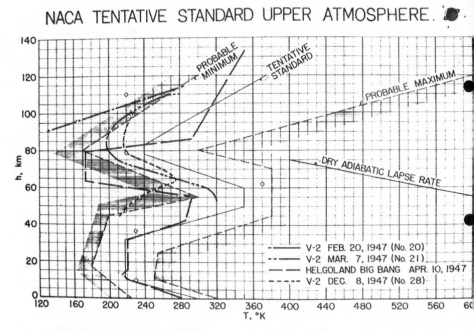

Figure 4.8
Standard Atmosphere to 120 kilometers as revealed by V-2 rocket flights, 1947. [*Helgoland Big Bang* refers to the acoustic signal produced by the detonation of 7 kilotons of conventional explosives in the North Sea]. *Source:* Wexler Papers 32.

new layers being discovered. Terms such as *turbosphere* (today's troposphere) and *suprasphere* (now the mesosphere) were open to negotiation. Names such as the *neutrosphere* (the opposite of ionosphere) never caught on.[94]

As mentioned in chapter 3, the first standard atmosphere, published by Gregg for the National Advisory Committee for Aeronautics in 1922, extended no higher than 20 kilometers, but by 1947 instrumented V-2 rockets had routinely reached six times higher uncovering new features of the atmosphere, new possible dynamical connections between the layers, and, most important, new environments in which missiles were expected to perform predictably (figure 4.8). The standard atmosphere of 1962 reached as high as 700 kilometers, with the region from 32 to 100 kilometers designated tentative and the region above 100 kilometers marked speculative.[95] It was a much bigger "ocean of air" than that contemplated by Torricelli in 1644 or by Vilhelm Bjerknes in 1904.

Chief Scientist for the International Geophysical Year

On April 27, 1955, Joseph Kaplan, chair of the U.S. National Committee, summoned Harry Wexler to a meeting at the International Geophysical Year headquarters for an "important" matter. "Upon arriving," Wexler noted in his diary, "I found Larry Gould, Lloyd Berkner, Joe Kaplan, Lincoln Washburn, Wally Joyce, and Hugh Odishaw. They asked me if I would accept the post of 'Chief Scientist' of the US/IGY Antarctic Expedition— duties to begin immediately."[96] Wexler was intrigued yet concerned about the effect that this assignment might have on his own research and responsibilities, at the time involving operational numerical weather prediction, general circulation modeling, scientific rocketry, and a host of other projects.

In the gray dawn of the following day, after tossing and turning most of the night, Wexler asked himself, "What am I getting into? Am I losing five years of my scientific life with global meteorological problems to concentrate on a small area to act as father confessor to frustrated Antarctic scientists—to battle with the Navy for scientific program priorities?" Wexler didn't agonize for long. That day, he consulted with polar meteorologist Paul Humphrey, who gave him some "useful background about Antarctic cultists," and then spoke to von Neumann about the computer project. After calling Reichelderfer, Wexler accepted the IGY position on the condition that he find fieldwork as enjoyable and productive as his desk and lecture duties. He had to convince himself that studying Antarctica's weather and climate, the Southern Hemisphere's oceanic and atmospheric circulation, and the interaction between the hemispheres held the keys to a better understanding of global meteorology. Wexler soon realized, however, that the moniker "chief scientist" was nominal at best. He recorded his frustrations in a letter to Reichelferfer:

If the Antarctic Chief Scientist job is mainly operations (as indeed it is), this should be recognized, the title changed, and an operational type of individual obtained. I am getting rather fed up with helping IGY on this "gratis" basis—at what seems to me at considerable sacrifice to the Weather Bureau's scientific program, as well as my commitments to other agencies.[97]

Six months had passed, and his appointment had not yet been officially announced, further hobbling Wexler and leading him to suggest that it would be easy to withdraw from the position. He wrote to Joseph Kaplan to tender his official resignation—to an unofficial appointment.[98] Yet the IGY leadership prevailed on him, and he soldiered on.

Wexler's Antarctic diary documents his struggles with the assignment and the contrast between the hectic pace of the International Geophysical Year and the rest of his life, his family, and his research. One entry reads, "15 Aug. 1956, 8 p.m. End of a perfect Sharon day—kite flying, Mrs. Burton's garden, and swim." Five days later, he noted that Odishaw and Gould had been calling several times each day, "badgering" him about the IGY.[99] After spending all of October 18 at the IGY headquarters, he wrote in his diary at 7:15 p.m., "A frantically fantastically busy day." He was off to MIT two days later, "and then unexpectedly at noon [driving] to Woods Hole with Charney, Phillips, Lorenz, and Kuo" because Rossby was there and he wanted to see them. A lecture by Ed Lorenz "made New Zealand seem far away and [led] me again to wonder why I'm in the midst of [the International Geophysical Year] and not working on meteorology."[100] A partial answer to his reverie came from a discussion led by Rossby and Charney after the talk concerning the role of land and ocean contrasts in the Northern Hemisphere in initiating blocking and whether the phenomenon would still exist in the Southern Hemisphere. Wexler was going to be in the Southern Hemisphere for several months and perhaps could find out.

Finding a suitable interim replacement for Wexler at the Weather Bureau was not an easy task. In December, he informed Reichelderfer that someone would have to replace him on the Joint Numerical Weather Prediction ad hoc committee. He was also concerned about reorganizing the work of the Special Projects Unit on atomic energy and air pollution; supporting the Mauna Loa Observatory; dealing with upcoming reports on meteorological training, evaporation suppression, melting acceleration, and radiation instrumentation; and addressing personnel matters, including promotions. Wexler wanted to return from the trip in time for the next meeting of the National Academy Committee on Meteorology scheduled for March 25–27, 1957, because they were deliberating the creation of a National Institute for Atmospheric Research (chapter 5). An important consideration was family. Harry and Hannah decided it would be too difficult if he were away for such a long time, especially over the holidays. Wexler decided he could perform his duties by flying to New Zealand and Antarctica in January and thus be away for only two months.[101] Rossby was not at all pleased about

Wexler's upcoming trips and told Wexler, "he had too many fingers in too many pies."[102]

In 1956, about the same time Rossby was preparing his article on "Current Problems in Meteorology," Wexler composed his own review paper on "Meteorology in the International Geophysical Year."[103] The two articles share many things in common, especially their broad perspectives, acute observations, and authoritative tone. Wexler began his essay with a global view of the rotating Earth—covered by a thin layer of atmosphere "proportionately as thin as the skin of an orange" and moving through space bombarded by cosmic rays, meteors, radiation from the stars, and energetic particles from the sun. Wexler explained how, at the uneven surface of the planet, the atmosphere absorbs heat, transports water vapor and other materials, and moves, opposed by friction, over land and water. Earth repays its debt to the sun and space by transferring energy upward in the form of infrared radiation. Wexler painted a picture of the atmosphere that was never at rest, driven by a great heat engine that generates a cascade of motions, from thousand-mile-long trade wind belts, westerly winds in mid-latitudes, planetary waves aloft, cyclones, hurricanes, thunderstorms, tornadoes, dust devils, and tiny vortices that tie it all together. Earth's axial tilt means that tropical and subtropical regions receive a surplus of solar radiation, while middle-latitude and polar regions run an annual energy deficit. As a consequence, air warmed and charged with water vapor at lower latitudes moves poleward, driving a global circulation system that generates all the weather and shapes all the climatic zones on the planet.

Wexler pointed out how humans have always been dependent on and often vulnerable to the vicissitudes of the weather and have taken steps to seek both protection and prevision. Forecasting was based initially on local observations, tradition, and weather lore and then on observations that extended far beyond the horizon and high into the sky by the use of telegraphy, radio, kites, balloons, airplanes, and rockets. Even so, Wexler admitted, only about one-fifth of the total atmosphere is adequately probed each day, with the vast oceans, the Southern Hemisphere, jungles, deserts, and the polar regions having the poorest coverage. Weather forecasts are not possible in regions with no or poor observations, but, as the Bergen school demonstrated for western Norway, useful short-term forecasts are possible where observations are more numerous. This is because, for a day or two, the local weather is influenced by regional conditions. According to Wexler, longer-range forecasts, still quite elusive, would require hemispheric or even global observations, but even those might not be sufficient. He had high hopes for the upcoming International Geophysical Year of 1957–58 and was inspired by how meteorologists have been wholeheartedly supportive of earlier observational campaigns such as the First Polar Year in 1882–83 and the Second Polar Year in 1932–33.[104]

The IGY was designed to extend existing meteorological observing networks laterally and vertically into the unknown. Wexler was involved in planning measurements of the annual global circulation and poleward transport patterns of energy, momentum, and water vapor, all resulting from daily and seasonal changes in the weather. According to the Bergen school, the main contributors to this circulation are horizontal eddies in the form of cyclones and anticyclones and vertical eddies such as the Hadley cell from the equator to 30 degrees latitude and the Ferrel cell in midlatitudes. Wexler relished the opportunity, for the first time, to compare circulation and transport patterns in the mostly maritime Southern Hemisphere with those in the more continental Northern Hemisphere and to look for worldwide feedback mechanisms, especially across the equator.

Wexler pointed out that meteorology plays a compelling role in other geophysical sciences from the study of the oceans to the top of the atmosphere. Indeed, the interplay between the atmosphere and ocean is so intimate that geophysical hydrodynamics treats it as one medium separated by an internal discontinuity in density and velocity. Important exchanges of heat, moisture, momentum, dissolved gases, salts, and other elements occur across the boundary known as the sea surface. Marine cloud systems ranging from ocean fogs and trade-wind cumulus to violent tropical storms are visible markers of the interaction of meteorological and oceanic conditions. In the upper level of the ocean, wind-driven currents carry significant amounts of heat poleward annually, while on time scales of decades, centuries, and millennia, immense pools of cold bottom water form and overturn, with large-scale effects on Earth's climate.

Wexler then mused about the higher levels of the atmosphere. The lower layer of weather is capped at altitudes between 10 and 15 kilometers by the stratosphere, a dry, stable region heated by absorption of solar ultraviolet radiation by ozone molecules. At the edge of space, energetic solar radiation and particles ejected by the sun impinge on Earth's thin atmosphere, causing aurora, airglow, and other electrical and geomagnetic phenomena.

Whether the vast changes experienced in this ionized region were due to solar flares and whether cosmic rays penetrating deep into the atmosphere influenced the weather below remained open questions. In general, observations planned for the International Geophysical Year were bound to illuminate the complex interplay of meteorology and related geophysical sciences.

Wexler, who was well versed in the history of his field, pointed out that the First and Second Polar Years had focused on the Arctic, while the International Geophysical Year was devoted to Antarctic geophysics in general, Antarctic meteorology in particular, and the open question of the sun's influence on the planet. There had never been more than four or five complete meteorological stations on the Antarctic continent, an area twice as large as the United States, but under Wexler's leadership, this number was set to increase by nearly fivefold. Understanding the extreme weather conditions in Antarctica, not found elsewhere on Earth, would fill an immense gap in meteorological knowledge. On Wexler's list of unique large-scale phenomena open for investigation were circumpolar circulation patterns around an open ocean, atmospheric vertical structure during the long polar night, high-speed drainage winds from the high central plateau, and the flux of subterranean heat through the massive ice sheet. Antarctica promised to provide an incomparable natural laboratory for studies of atmospheric energy exchange and turbulence over the flat snow surface of the Ross Ice Shelf. Local measurements of ozone, carbon dioxide, and radioactivity would be compared with those taken elsewhere.

Wexler was deeply involved in the planning of Earth-orbiting satellites to study the global heat balance and help discern long-term trends of world climate. If a 1 percent excess of energy, for example, was found to be entering the system, it would have to go somewhere—toward an imperceptible heating of the oceans, a hard-to-detect melting of glacial ice, a dramatic warming of the atmosphere, or more likely, a combination of all three. He was concerned by the use of the atmosphere as a dumping ground for pollutants, both locally and in a global sense. He warned that the burning of fossil fuels could result in a doubling of carbon dioxide concentrations in the next half century, while smoke, ash, and other emissions could reduce solar radiation at the surface and interfere with natural precipitation processes. Global heating or cooling could be the result of this massive but uncontrolled "experiment." He believed that nuclear contamination of the air, water, and soil was also an issue and that a new epoch of global monitoring of atmospheric composition was clearly needed. Wexler was in charge of Mauna Loa Observatory and gave it a special measure of his interest and attention, visiting it on his trip to New Zealand and Antarctica. He succeeded in providing them with an infrared gas analyzer, "to keep a continuous record of CO_2 at the Observatory." Writing in 1955, well before the acknowledged dawn of the environmental movement, Wexler placed an additional heavy burden of responsibility on the world's geoscientists to ensure that basic environmental measurements, crucial to life as we know it and to the health of future generations, are carefully recorded and that new ones are initiated and maintained. This was a new moral responsibility in science.[105]

Weather Modification

Irving Langmuir, a Nobel laureate in surface chemistry but untrained in meteorology, routinely exaggerated the possibilities of weather and climate control. He claimed the cloud-seeding techniques that his team had recently developed at the General Electric Corporation could be used to redirect hurricanes, suppress lightning strikes, put out forest fires, eliminate aircraft icing and fog hazards, and make it rain in arid environments. He offended meteorologists when he told them they might as well give up forecasting because a drop of water or a few grains of dry ice when properly implanted in the atmosphere will result in significant storms. He also made the outrageous claim that he could control the mushroom cloud of an atom bomb by salting it with a few dashes of silver iodide.[106] The strategic implications of weather control led to further studies by the military in Project Cirrus from 1947 to 1952; in the Cloud Physics Project of the Weather Bureau, air force, and National Advisory Committee for Aeronautics from 1948 to 1951; and in the military's Artificial Cloud Nucleation Project from 1952 to 1957. The case studies were inconclusive.

Reichelderfer advised Wexler and his associates to avoid "persiflagery" and to keep an open mind on weather control because the Weather Bureau had received federal funding to study the subject. It was impossible to resist; Colonel Ben Holzman responded, "I still think that a good umbrella is a very reliable weather control gadget."[107] Wexler sought to protect the public from commercial rainmakers. He pushed back against their excessive claims (writing, for example, a letter to the editor of *Time* magazine) and sought a strong statement against such practices from the American Meteorological Society.[108] Reichelderfer too was passionate about Langmuir's disservice to the citizens of the United States. He wrote in 1950:

It seems to me it is time we began to play up authentic meteorological opinion in this matter of rainmaking in view of the complete abandon with which the proponents of artificial rainmaking on a large scale have expressed themselves in public over the past two or three years without much regard for scientific evidence or the pocketbooks of the poor farmers who pay for rain which they think the commercial rainmakers in the Southwest are making for them. I do not say that no rain can be "induced" by artificial means but I do say emphatically that relatively little of the rain increase claimed by commercial rainmakers has actually been due to artificial factors.[109]

In 1953, during the height of cloud-seeding activities, approximately $5 million a year was being spent on commercial cloud-seeding operations over about 10 percent of the continental United States. Irving Krick ran one of the largest companies. In that year, Congress established the President's Advisory Committee on Weather Control to study and reevaluate the situation. Chaired by retired navy captain Howard T. Orville, the committee's final report lent some credibility to the practice of commercial cloud seeding. Orville flirted with professional censure by the American Meteorological Society, however, when he authored a popular article in *Colliers* hyping the potential of weather control as a weapon of war.[110]

By 1955 Reichelderfer thought almost a decade had been wasted by premature exploitation, and if the Bureau's recommendations had been followed, "we would have got down to fundamentals in the first place and be well on the way to *knowing* what the possibilities are. As it is, the fanatics have so criticized the Weather Bureau for keeping its feet on the ground that we have been unable to compete for research projects in this field against those who promise the moon!" The weather-control situation had evolved far beyond discussion of chemical seeding agents, however. Edward Teller proposed the use of nuclear weapons to intervene in storms, John von Neumann thought digital computers might provide the means to examine the consequences of future climate interventions, and nascent space scientists imagined the capabilities of Earth-orbiting satellites for observing and controlling the weather.[111]

Wexler drove these points home when he addressed the National Rivers and Harbors Congress on the possibility of weather control and its implications for the nation's water supply. Satellites certainly could be used to monitor rivers, reservoirs, and coastlines during times of drought or flood. He turned then to the dangers of weather modification because of the variables and unknowns of the atmosphere: "With the growing energy resources in hand there will be a strong temptation to perform [weather-control] experiments and perhaps bring about an untenable situation. It is even more important in this larger area that we know what we are doing and in this respect the satellite with its increased data gathering capabilities and the hi-speed electronic computer, with its lightning fast calculations, will be indispensable tools for the meteorologist."[112]

Meteorological Satellites

Space technology affords new opportunities for scientific observation and experiment, which will add to our knowledge and understanding of the earth, the solar system, and the universe.
—President's Science Advisory Committee, 1958[113]

Encouraged by novelist and futurist Arthur C. Clarke, Wexler explored the possibilities of Earth observations from space in his 1954 lecture presented at a symposium on space travel held at the Hayden Planetarium in New York City.[114] For the occasion, he commissioned an artist's impression of Earth from space in full color, showing clouds, land, and ocean and depicting weather features, such as a family of three cyclonic storms along the polar front, a small hurricane embedded in the trade winds, evidence of jet stream winds, and fog off the coasts (figure 4.9).[115]

Wexler published versions of his remarks in the *Journal of the British Interplanetary Society* and the *Journal of Astronautics*, where he made strong claims for the utility of the meteorological satellite, not only as a "storm patrol" but also as a potentially revolutionary new tool with global capabilities:

Since the satellite will be the first vehicle contrived by man which will be entirely out of the influence of weather it may at first glance appear rather startling that this same vehicle will introduce a revolutionary chapter in meteorological science—not only by improving global weather observing and forecasting, but by providing a better understanding of the atmosphere and its ways. There are many things that meteorologists do not know about the atmosphere, but one thing they are sure of is

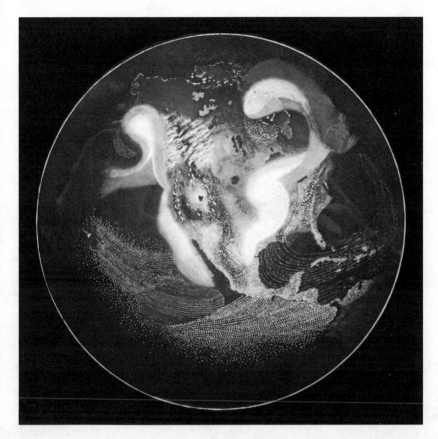

Figure 4.9
Artist's impression of Earth from space, 1954. Original caption: "This is an imaginary
weather picture taken from a rocket 4,000 miles above Texas at noon June 21. Vast
cloud masses associated with three storms in different stages of development cover
the mid-western United States and Hudson Bay area. A zone of thunderstorms and
tornadoes stretches from Cleveland to New Orleans. Cumulus clouds borne by trade
winds stream westward across the Atlantic and Pacific. A hurricane is located north
of Puerto Rico." The painting, commissioned by Harry Wexler in 1954, hangs in the
conference room of the National Environmental Satellite, Data, and Information
System at the National Oceanic and Atmospheric Administration in Silver Spring,
Maryland. *Source:* Wexler Papers 41; color image bit.ly/406earthfromspace1954.

this—that the atmosphere is indivisible—that meteorological events occurring far away will ultimately affect local weather. This global aspect of meteorology lends itself admirably to an observation platform of truly global capability—the Earth satellite.[116]

In 1955, the United States announced it would launch a scientific Earth satellite during the International Geophysical Year of 1957–58. To promote the program, Weather Bureau officials lectured on the remote measurements that would be possible with meteorological satellites. These included retrieval of temperatures from the stratosphere, tropopause, cloud tops, and the surface; identification and measurement of trace constituents such as water vapor, ozone, and carbon dioxide; information on Earth's overall heat budget; and the ability to track clouds, precipitation layers, and severe storms (figure 4.10).

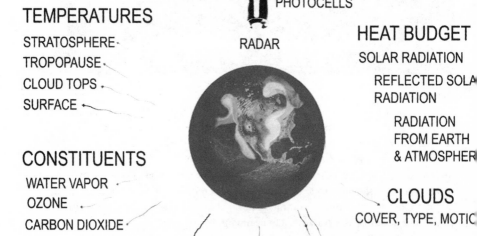

MEASUREMENTS WITH METEOROLOGICAL SATELLITES

TELEVISION — IR DETECTORS / PHOTOCELLS

TEMPERATURES
STRATOSPHERE
TROPOPAUSE
CLOUD TOPS
SURFACE

RADAR

HEAT BUDGET
SOLAR RADIATION
REFLECTED SOLAR RADIATION
RADIATION FROM EARTH & ATMOSPHERE

CONSTITUENTS
WATER VAPOR
OZONE
CARBON DIOXIDE

CLOUDS
COVER, TYPE, MOTION

THUNDERSTORMS   TROPICAL STORMS

HEIGHT LAYERS / PRECIPITATION } RADAR

Figure 4.10
Slide illustrating anticipated measurements using meteorological satellites, circa 1955. The central image is Wexler's commissioned painting, depicted in figure 4.9.
Source: NARA, RG 370-MSP SER 2-B.

The launch of the Soviet satellite *Sputnik 1* on October 4, 1957 diverted the world's attention from scientific concerns and focused American perceptions on a "missile gap" and possible national security threats from space. Less than one month later, *Sputnik 2* further fueled these fears. In response and with the U.S. Navy's Vanguard Program languishing, the U.S. Department of Defense used a modified Redstone military missile, the Juno 1, to launch the first U.S. satellite, *Explorer 1*, on January 31, 1958. In that year, Congress enacted the National Defense Education Act, which provided dramatically increased support for both basic research and science education at all levels. The National Aeronautics and Space Act created a new agency—the National Aeronautics and Space Administration—to consolidate and lead the U.S. space effort. Its mission included expanding knowledge of phenomena both within the atmosphere and in outer space and developing and operating vehicles capable of carrying instruments for peaceful and scientific purposes in cooperation with other nations. Also in 1958, the President's Science Advisory Committee pointed out that a satellite in orbit could be used for three scientific purposes: "(1) it can sample the strange new environment though which it moves; (2) it can look down and see the earth as it has never been seen before; and (3) it can look out into the universe and record information that can never reach the earth's surface because of the intervening atmosphere."[117]

On May 1, 1958, University of Iowa scientist James Van Allen announced that Geiger-Müller counters aboard the Jet Propulsion Laboratory's *Explorer 1* and *Explorer 3* satellites had been swamped by high radiation levels at certain points in their orbits, indicating that powerful radiation belts, later known as the Van Allen belts, surround Earth. *Vanguard 1*, the fourth artificial satellite launched, provided important geodetic information about the shape of Earth, specifically its north-south asymmetry.[118] *Vanguard 2* used photocells to measure sunlight reflected from clouds, and *Explorer 6* photographed them. Van Allen was deeply involved in developing space science instrumentation. He and Verner Suomi designed a flat plate radiometer to measure Earth's heat balance that flew on *Explorer 7* in 1959.[119]

On September 10, 1958 Wexler called his staff together to inform them that, "the Weather Bureau is about to enter the Space Age." He had just received notification that the Weather Bureau had been designated as NASA's meteorological agent to provide instrumentation, data reduction, and analysis of observations taken by satellites after the International

Geophysical Year ended.[120] The meteorological satellite era officially began on April 1, 1960, when *TIROS 1* (the first television and infra-red observing satellite) reached orbit. It was a collaborative project of NASA, the U.S. Army Signal Research and Development Lab, the RCA Corporation, the U.S. Naval Photographic Interpretation Center, and the U.S. Weather Bureau. Wexler served as the chief scientist for the *TIROS* program. *TIROS 1* provided the first experimental test of the utility and research potential of a weather satellite. It carried two shuttered television cameras (photosensitive vidicon tubes) that recorded images of clouds on tape for later transmission to the ground.

On April 1, 1960, Wexler participated in the first *TIROS 1* press conference. NASA chief administrator T. Keith Glennan set a celebratory tone, emphasizing that the launch was a "total national effort" involving several agencies and industries, by implication, the Congress and American public. Abe Silverstein, NASA's space flight development director, then provided some of the technical details of the satellite, launch, and orbit. When asked about early results, Morris Tepper, head of the meteorological satellite program at NASA, admitted that they did not have any yet, at least for distribution. When asked if the satellite was going to improve the accuracy of weather forecasts, Wexler replied in the affirmative, if only because the satellite will fly over meteorological blind spots—areas with no current observations, estimated to be 80 percent of the planet. Although it was too early to provide details, the satellite would also provide a new way of measuring basic physical parameters, so the prospects were looking good.[121]

Three weeks later, *TIROS 1* had completed over three hundred orbits and returned some seven thousand cloud photographs. At a press conference on April 22, Reichelderfer expressed his hope that a new era of meteorological observing was opening up that would provide hitherto missing observational data for researchers and better support for the forecasting service. He praised Wexler, who was sitting next to him, for having "the gleam in his eye about what satellites might do long before the rest of us did" and for providing "an excellent forecast, six-years in advance." He was referring to Wexler's 1954 talk at the Hayden Planetarium and the artist's image he commissioned of a satellite view of Earth (figure 4.9). Wexler then shared his favorite geophysical joke: "Oceanographers have a saying that seventy percent of the world is covered by water; meteorologists, that *one hundred percent* of the earth is covered by air." They now had an observing platform with global implications. Initial results included detection of a potential tornadic area over Nebraska, a tropical cyclone in the South Pacific east of Australia, and a massive storm 1,300 kilometers west of California with banded structures never before seen by meteorologists. Wexler provided the first-ever really cosmic weather map discussion. He displayed sequences of cloud images with superimposed analyzed charts depicting Bergen school fronts and streamlines (figure 4.11). Wexler wanted to use *TIROS* images to determine the percentage of tropical storms with banded structure and whether more occur over the oceans. When asked if *TIROS* would help in extending long-range forecasts, Wexler replied that any progress along these lines would need the kind of global observations that the satellite was providing. The weather satellite was not replacing other observations but was actively supplementing them—for example, by locating offshore storms so that reconnaissance aircraft flights could be planned more efficiently.[122]

TIROS 1 demonstrated the viability of a space-based weather patrol. It had a spin-stabilized space-oriented (not ground-oriented) orbit, daylight only sensors, and a short seventy-eight-day lifetime, which limited its utility, but it proved the concept, and meteorologists found the images both inspirational and useful in their work. *TIROS 2* (November 12, 1960) and *TIROS 3* (July 12, 1961) were equipped with infrared sensors for attitude control and shortwave and longwave radiation experiments. *TIROS 3*, launched in hurricane season, provided photographs of Hurricane Esther, the first tropical cyclone to be discovered by satellite imagery. *TIROS 4* (February 8, 1962) was outfitted with a new lens system and provided the best pictures to date, good enough for the Weather Bureau to share internationally over its facsimile transmission system. *TIROS 5* (June 19, 1962) reached near operational capability during the hurricane season. Due to a launch anomaly, its orbit facilitated observations at higher latitudes than its predecessors, including clear photographs of the summer breakup of Arctic sea ice.[123] The satellites served as "storm patrols" for early warnings, as aids to weather analysis and forecasting, and as research tools for atmospheric scientists. They provided panoptic images never seen before: tropical storms approaching landfall in remote areas, mountain wave clouds over the Andes, and circumpolar circulation patterns in both hemispheres (figure 4.12). Wexler explained that "the TIROS satellites disclosed the existence of storms in areas where few or no observations previously existed, revealed

Figure 4.11
Wexler inspecting *TIROS* photographs and weather maps in 1960. Photo shoot by *Life* magazine photographer Edward Clark staged at home in the girls' bedroom.

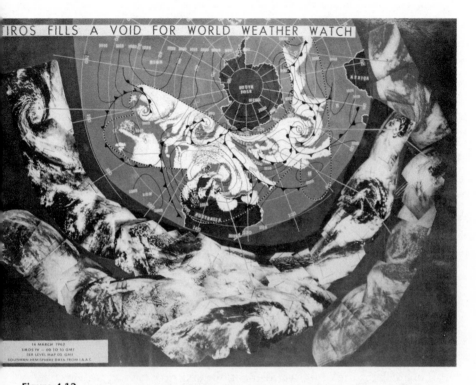

Figure 4.12
"*TIROS* fills a void for World Weather Watch." Map of a portion of the Southern
Hemisphere with cloud pictures taken by *TIROS 4* superimposed on a conventional
contour analysis of surface conditions on March 14, 1962, as provided by the International Antarctic Analysis Centre. *Source:* NARA, Reichelderfer Records, box 4.

unsuspected structures of storms even in areas of extensive observational
coverage, depicted snow fields over land, ice floes over water, and temperature patterns on land and ocean as well as temperatures of tops of cloud
layers." Because its accomplishments were clearly accessible to the general
public, the program enjoyed strong political support, demonstrating that
sustained weather observations from space were possible.[124]

Climate Engineering

In 1958, Wexler published an article that examined some of the consequences of tinkering with Earth's heat budget. He began by describing
the two streams of radiant energy and their seasonal and geographic

distribution—one stream directed downward and the other upward, which "dominate the climate and weather of the planet Earth." Gains consist of the solar radiation absorbed by Earth's surface and atmosphere after accounting for losses by reflection. Energy losses involve infrared radiation emitted to space by Earth's surface and atmosphere, the latter mostly from atmospheric water vapor, clouds, carbon dioxide, and ozone. Wexler wrote: "In seeking to modify climate and weather on a grand scale it is tempting to speculate about ways to change the shape of these basic radiation curves by artificial means," especially by changing the reflectivity of the earth. He was opposed to using carbon dust to blacken the deserts and the polar ice caps. Even more outrageous was the notion, probably originating with Edward Teller, that detonating ten really "clean" hydrogen bombs in the Arctic would produce a dense ice cloud at high latitudes and would likely result in the removal of the sea ice. The balance of Wexler's paper examines the radiative, thermal, and meteorological consequences of this outrageous act. He noted, perceptively, that "the disappearance of the Arctic ice pack would not necessarily be a blessing to mankind" and implied that the United States and the Soviet Union already had the firepower to try such an experiment. Wexler had heard many less than serious proposals on the climatological use of nuclear detonations and may have at least wondered what effect a nuclear war would have on climate. He concluded his essay with an insight whose relevance has not been diminished by time:

When serious proposals for large-scale weather modification are advanced, as they inevitably will be, the full resources of general-circulation knowledge and computational meteorology must be brought to bear in predicting the results so as to avoid the unhappy situation of the cure being worse than the ailment.[125]

In 1962, informed by the latest results from computer climate models and satellite measurements of Earth's heat budget, Wexler expanded his study to examine theoretical questions concerning natural and anthropogenic climate forcings, both inadvertent and purposeful. He lectured "On the Possibilities of Climate Control" at the Boston chapter of the American Meteorological Society, at the Traveler's Research Corporation in Hartford, and at the UCLA Department of Meteorology. Wexler reminded his listeners that recent high-level diplomatic initiatives by U.S. president John F. Kennedy and Soviet premier Nikita Khrushchev had made climate modification "respectable" to talk about. He then quoted extensively from Zworykin's 1945 weather control proposal and von Neumann's response to it.[126]

Wexler was clearly interested in both inadvertent climatic effects—such as might be created by industrial emissions, rocket exhaust gases, or widespread land use changes—and purposeful climate interventions, whether peaceful or done with hostile intent. He discussed the problem of increasing global pollution from industry, citing rising carbon dioxide emissions as an example of indirect control. He mentioned the Callendar effect as one of the ways that humanity was already inadvertently modifying global climate: "We are releasing huge quantities of carbon dioxide and other gases and particles into the lower atmosphere, which may have serious effects on the radiation or heat balance, which determines our present pattern of climate and weather."[127] He then reviewed recent technical developments that led him to conclude that purposeful interventions in global atmospheric processes would soon be possible if not yet probable. Wexler was concerned with planetary-scale manipulation of the environment that would result in "rather large-scale effects on general circulation patterns in short or longer periods, even approaching that of climatic change." He assured his audiences that he did not intend to cover all possibilities "but just a few . . . *limited primarily to interferences with the Earth's radiative balance on a rather large scale*: I shall discuss in a purely hypothetical framework those atmospheric influences that man might attempt deliberately to exert and also those which he may now be performing or will soon be performing, perhaps in ignorance of its consequences. We are in weather control *now* whether we know it or not."[128]

In 1955, in a prominent article in *Fortune* magazine, von Neumann had asked, provocatively, "Can We Survive Technology?" He thought that weather control using chemical agents and climate control through changing surface albedo or otherwise modifying solar radiation were distinct possibilities for the near future, but he called it a thoroughly "abnormal" industry. He argued that such intervention could have "rather fantastic effects" on a scale difficult to imagine and pointed out that altering the climate of specific regions or purposely triggering a new ice age was not necessarily a rational undertaking. Tinkering with the earth's heat budget or the atmosphere's general circulation "will merge each nation's affairs with those of every other more thoroughly than the threat of a nuclear or any other war may already have done." In his opinion, climate control, like other "intrinsically useful" modern technologies, could lend itself to unprecedented destruction and to forms of warfare as yet unimagined.[129]

Echoing von Neumann, Wexler warned that our technological leverage over the planet might exceed our wisdom: "Even in this day of global experiments, such as the world-wide Argus electron seeding of the Earth's magnetic field at 300 miles height, man and machinery orbiting the Earth at 100 miles seventeen times in one day, and 100 megaton bombs—are we any closer to some idea of the approaches which could lead to an eventual 'solution' [to the problem of climate control]?" He noted "a growing anxiety" in the public pronouncements that "Man, in applying his growing energies and facilities against the power of the winds and storms, may do so with more enthusiasm than knowledge and so cause more harm than good." Wexler was well aware that any intervention in the earth's heat budget would change the atmospheric circulation patterns, the storm tracks, and the weather itself and that, as he pointed out, weather and climate control are not two different things.[130]

After presenting some twenty technical slides on the atmosphere's radiative heat budget and discussing means of manipulating it, Wexler concluded his lecture with a grand summary of newly available techniques to heat, cool, or otherwise restructure the atmosphere: (1) increasing global temperature by 1.7 degree Centigrade by injecting a cloud of ice crystals into the polar atmosphere by detonating ten hydrogen bombs in the Arctic Ocean; (2) lowering global temperature by 1.2 degrees Centigrade by launching a ring of dust particles into equatorial orbit to shade the earth; (3) warming the lower atmosphere and cooling the stratosphere by injecting ice, water, or other substances into space; and (4) destroying all stratospheric ozone, raising the tropopause, and cooling the stratosphere by up to 80 degrees Centigrade by an injection of a catalytic deozonizing agent such as chlorine or bromine. Wexler warned that the space age was introducing an entirely new kind of "atmospheric pollution" problem. He was particularly worried that some types of rocket fuel, when burned, might release chlorine or bromine, "which could destroy naturally occurring atmospheric ozone and open up a 'hole,' admitting passage of harmful ultra-violet radiation to the lower atmosphere." Changes in the upper atmosphere caused by increasing jet contrails, space experiments gone awry, or the actions of a hostile power could disrupt the ozonosphere, the ionosphere, or even the general circulation and climate on which human existence depends. Wexler felt that it was urgent to use "the most advanced mathematical models of atmospheric behavior" to study the physical, chemical, and meteorological consequences of such interferences. He then explained how the Weather Bureau was in the process of acquiring new computers and developing new models to "simulate the behavior of the actual atmosphere and examine artificial influences that Man is introducing in greater and greater measure as he contaminates the atmosphere."[131]

Destroying stratospheric ozone was one of the most stunning aspects of Wexler's lectures. He was aware that catalytic reactions of chlorine and bromine could severely damage the ozone layer. He was concerned that increased rocket exhaust could pollute the stratosphere and that near-space "seeding" experiments might have unintended consequences. He also feared that the Cold War and the space age might provide rival militaries with both the motivation and wherewithal to inflict purposeful damage on the ozone layer. He cited a 1961 study by the Geophysics Corporation of America on possible harm to Earth's upper atmosphere caused by the oxidizers in rocket fuel. He was also aware that operations Argus and Starfish (the detonation of nuclear bombs in the magnetosphere), Project West Ford (injecting copper needles into orbit), and Project High Water (dumping water into space just to see what it would do) constituted recent significant interventions in the near-space environment that were accompanied by unknown and unquantified risks.[132]

On the topic of purposeful damage, Wexler turned to the 1934 presidential address to the Royal Meteorological Society, in which the noted geoscientist Sydney Chapman had asked, "Can a hole be made in the ozone layer?" That is, can all or most of the ozone be removed from the column of air above some chosen area? Chapman was thinking of an event that would provide a window for astronomers to extend their observations some hundreds of ångströms farther into the ultraviolet without the interference of atmospheric ozone. Possible health effects of human exposure to short-wave radiation did not appear to Chapman to be an important issue because the hole he was contemplating would be localized, probably in a remote area (he suggested Chile), and would be short-lived (somewhere between a day and an hour), timed for the benefit of astronomers only. Cutting such a hole, Chapman continued, would require "the discharge of a deozonizing agent," perhaps by airplanes, balloons, or rockets. Two possibilities came to mind—a large amount of a one-to-one destructive agent such as hydrogen that would reduce O_3 molecules to O_2 or "some catalyst which, without itself undergoing permanent change, could promote the reduction of large numbers of ozone molecules in succession."[133] By removing the ozone layer from the radiation model of Syukuro Manabe and Fritz Möller, Wexler was able to calculate a catastrophic 80 degree Centigrade stratospheric cooling that would occur if Earth had no ozone layer. He wondered if such a technique could be weaponized.[134]

Seeking advice on how someone might cut a "hole" in the ozone layer, Wexler turned to chemist Oliver Wulf, now at Caltech, who suggested that "from a purely chemical viewpoint, chlorine or bromine might be a 'deozonizer.'" Wulf and Wexler exchanged numerous letters between December 1961 and April 1962 and met in March; Wulf met with Chapman in April. All these exchanges point to the conclusion (a stunning one, given the received history of ozone depletion) that chlorine or bromine atoms might act in a catalytic cycle with atomic oxygen to destroy thousands of ozone molecules. Wexler estimated that a 100-kiloton bromine "bomb" would destroy all ozone in the polar regions, and four times that amount would be needed near the equator. Essentially, Wexler and Wulf had identified the basis of the modern ozone-depleting chemical reactions. Prominent atmospheric physicist Chankey N. Touart, who had heard Wexler's lecture in Boston, wrote that climate engineering constituted "mental gymnastics. . . . Let's hope the entire world is satisfied to play the game on this plane until the state of meteorological knowledge is truly adequate for big league experimentation." Wexler replied, "I hope that before we get into large experimentation that not only will the state of meteorological knowledge be much more advanced than it is now, but also the state of our sociopolitical affairs as well."[135]

Wexler and the World Weather Watch

In view of the rapid population rise and growing necessity to utilize efficiently and without waste all of the world's resources dependent on weather, *the human race can no longer afford to leave unknown any significant portion of the world's atmosphere.*
—Wexler, 1962[136]

January 1961 marked a new age for international cooperation in science. In his inaugural address on January 20, President John F. Kennedy urged that the United States and its adversaries "seek to invoke the wonders of science instead of its terrors." Ten days later, in his State of the Union address,

Kennedy made explicit his intention "to explore promptly all possible areas of cooperation with the Soviet Union and other nations," specifically in weather-prediction, satellites, and interplanetary probes. In February 1961, Wexler received an urgent call to travel to American Meteorological Society headquarters in Boston for a meeting on international cooperation in meteorology in support of President Kennedy's recent messages.[137] Joining Wexler were Jule Charney, Henry Houghton, Sverre Petterssen, Morris Tepper, atmospheric physicist Richard Goody, satellite meteorologist David Johnson, and Thomas Malone from the National Academy of Sciences Committee on Atmospheric Sciences (chapter 5).

The group proposed a program involving many nations in varying degrees to study global atmospheric processes, develop the scientific basis of weather prediction, and investigate the feasibility of large-scale modification of weather and climate. Their technical recommendations included using the new observational tools of satellites and rocketsondes to supplement surface and upper-air observations and to provide measurements over inaccessible regions; using mathematical modeling of the atmosphere and the most advanced electronic computers to analyze and simulate the physical processes that produce weather over a broad spectrum of scales; applying automation to data processing, analysis, and communication; and using field experimentation to explore the possibilities of artificially altering normal atmospheric behavior.

The working group produced a memorandum, "Possibilities in Meteorology for International Cooperation in Science," that identified why meteorology constituted the optimal vehicle for international scientific exchange and international bridge building: (1) atmospheric motions cross and recross international boundaries and even hemispheres; (2) it is scientifically untenable to isolate and study one portion of the atmosphere from the whole; (3) weather phenomena affect everyone everywhere; (4) collaboration aimed at broad, peaceful, humanitarian objectives might well stir the imagination of people in many nations; (5) a substantial amount of international exchange exists already in meteorology; (6) many developing countries need help in conducting an adequate weather service; (7) a shortage of scientific manpower exists, even in leading nations; (8) large economic benefits might be expected from advances in forecast accuracy; and (9) chances for success in large-scale control are possible but not definite. The next steps included informal deliberations with Soviet scientists to gauge their reactions to the scientific merits of the proposal, consultations with meteorologists here and in other countries, and a final recommendation to the president. This working group and their report essentially initiated the process, led by Petterssen, of charting the course of atmospheric science for the next decade (chapter 5).[138]

On March 1, 1961, at his fifth press conference, President Kennedy noted that meteorology and possibly weather control would constitute areas in which the United States could cooperate with the Soviets without any harm to national security. To Wexler's delight, on May 25, in his famous "moon landing" address before a joint session of Congress, the president requested $53 million for the Weather Bureau "to help give us at the earliest possible time a satellite system for world-wide weather observation." Additionally, on September 25, in an address to the General Assembly of the United Nations, Kennedy proposed reserving outer space for peaceful purposes, and pursuing "cooperative efforts between all nations in weather prediction and eventually in weather control."[139] These speeches, the work of meteorologists behind the scenes, and several important diplomatic and technological achievements provided a boost to international cooperation in the atmospheric sciences and services. In December 1961, the UN General Assembly passed Resolution 1721, "International Co-operation in the Peaceful Uses of Outer Space," which recommended advancing "the state of atmospheric science and technology so as to provide greater knowledge of basic physical forces affecting climate and the possibility of large-scale weather modification, and developing weather forecasting capabilities internationally, with the World Meteorological Organization to take the lead in both programs."[140]

In January 1962, Arthur Davis, the secretary-general of the World Meteorological Organization, asked Wexler to serve as an adviser, along with his Soviet counterpart, Viktor A. Bugaev, on the preparation of a report called for by the UN Resolution. Wexler coordinated his response with the U.S. National Academy of Sciences and arrived at World Meteorological Organization headquarters in Geneva on January 21 to draft the report. Afterward, he stopped in Paris at the headquarters of the United Nations Educational, Scientific and Cultural Organization to expand the scope into related environmental fields of oceanography and hydrology. He also discussed possible threats of upper-atmosphere contamination from rocket exhausts. Wexler was a guest lecturer for the month of February 1962 at UCLA. There he discussed with Jule Charney how to coordinate the idea of a World Weather Watch with the National Academy's interest in a global atmospheric research program (chapter 5). Wexler also networked with other leading geoscientists, prepared a second draft of the UN report, and sent M. A. Alaka, a former WMO staff member and Weather Bureau employee, to Geneva for the next several months to assist in the negotiations. Wexler returned to Geneva in March, where he worked with Bugaev to draft the final report. They had high-level support. Even as they met, President Kennedy exchanged notes with Soviet Premier Khrushchev: "Perhaps we could render no greater service to mankind through our space programs than by the joint establishment of an early operational weather satellite system." Khrushchev replied, "It is difficult to overestimate the advantage that people would derive from the organisation of a world-wide weather observation service using artificial earth satellites. Precise and timely weather prediction would be still another important step on the path to man's subjugation of the forces of nature."[141]

Wexler's concerns about rocket exhaust damaging the atmosphere are symbolized in a photograph of him taken in Geneva with Bugaev. The photo (figure 4.13) has three foci—the rocket itself soaring upward, the friendly relationship of Wexler and Bugaev, and the more ominous rocket exhaust plume. Wexler wrote at the time, "The exhausts from increasingly powerful and numerous space rockets will soon be systematically seeding the thin upper atmosphere with large quantities of chemicals it has never possessed before or only in small quantities."[142] A heavy travel schedule took a heavy toll on Wexler's health. In the first week of April he returned to Geneva to meet with representatives of international environmental agencies and to approve the penultimate draft of an agreement. There Wexler outlined his ideas about the World Weather Watch concept and read a six-page prospectus on a global atmospheric research program prepared by Charney. The seeds were sown. On May 15, Wexler delivered his final report to U.S. officials for approval and took his third trip of the year to Geneva to attend more high-level consultations, including a bilateral U.S.-USSR meeting on cooperation in space. He returned to Washington on June 10, 1962, sick and exhausted but with yet more to do, including a trip to Iowa in early July to brief the National Academy's Space Science Summer Study Group on the new developments.[143] With a portmanteau full of documents, Wexler set off with his family on a working vacation in Woods Hole, where a

Figure 4.13
Harry Wexler and Viktor A. Bugaev in Geneva at the World Meteorology Organization, March 19, 1962. *Source:* Wexler Papers 23.

summer study program on geophysical fluid dynamics was underway. There he caught up with Arnt Eliassen and Henry Stommel and student fellows James Holton, Joseph Pedlosky, and Carl-Gustaf Rossby's son, Hans Thomas. The article he wrote that summer on the World Weather Watch, published in the August issue of *Weatherwise*, would be his last.[144]

Personalizing Harry Wexler

Harry Wexler was not well. He had been diagnosed as being susceptible to "temporary coronary insufficiencies brought on by extreme fatigue," a condition his doctor told him could cause "mild heart flare-ups."[145] Still he maintained a brutal schedule. He died on August 11, 1962, at age fifty-one, three weeks before the eighteenth birthday of his first daughter, Susan, while he and his family were in Woods Hole. Here are Susan's memories:

I was at a social function at the Oceanographic building and when I walked into the house at 62 Millfield [Street] by Eel Pond, my uncle Jerome and aunt Edith Namias told me (this is all by memory) that dad was sick upstairs in bed. They may have said he had a heart attack. I know that Dr. Hirshfeld was with him. Instead of taking him to a hospital right away, for some reason they waited until the morning to bring an ambulance around. He lay on the sofa while people milled about, and then he was taken away in the ambulance. I turned to my aunt Edith outside by the picket fence and told her I would never see him again. I was right.

What I remember about dad: lots of fun, playing football, swimming at Lake Massapoag in Sharon, Massachusetts, being taught to swim by him, a big man, great smile, patience galore when it came to untangling kite string [figure 4.14]. Loved flying kites in the hay fields next to our house and cabins in rural Sharon, loved bicycling on his English Raleigh bike on which he taught me to ride, animated, loved to talk about the weather, pointed out different types of clouds, very gregarious, social, made friends easily, spent a lot of time in the study upstairs in our house in Falls Church in the evening going over his research. . . . I had a desk in the L shaped study around the corner and would do my homework there. He helped me with the Plane Geometry, which I was terrible at, but I squeaked by thanks to him. He liked working in the backyard raking the leaves in the fall and carrying them in a big sheet to the bottom of the driveway. He loved being outdoors. He would spend summers in Sharon helping out, cutting wood, tarring the cabin roofs. He took me as a youngster to the Blue Hill Observatory. We walked to the top in the dry summer heat, and I had the best piece of blueberry pie I have ever eaten in my life.

I was a daddy's girl and loved his animated love of living. When he went on his trips to Antarctica the house grew very quiet. He always brought something back from his trips for us, which was exciting. He drove me every Saturday to modern dance classes in Georgetown. I would lay in the back seat of the big 1950's Chrysler, and we would both listen to an episode of *Gunsmoke* on the radio. We did this for five years.

I remember every summer he would pack the Chrysler to the top with our things and drive from Falls Church, Virginia to Sharon, Massachusetts while mom, Libby and myself would get on a plane and fly up there. While we waited in the gathering darkness at Lalla's big house in the country and with the whip-poor-wills calling, we would hear the car pull into the driveway and know that dad had made it safely through the long ten-hour drive, and we would run down to the cedar-shingled garage and help unload everything we would need for the next two months.

Somehow dad was able to take six to eight weeks off each summer and stay with us in Sharon. We would go swimming, eat meals together, and shoot arrows at the

Figure 4.14
Harry Wexler kite flying with daughters Susan and Libby, Sharon, Massachusetts.
Source: Wexler Family Papers.

canvas-covered round target sitting on mom's easel in the field. They had custom-made bows and arrows, the long kind, not the modern contraptions.

Dad's best friend Jerome Namias had married mom's sister, Edith Paipert, and these two couples were very close until dad's death. Jerome, Edith and my cousin Judy lived in an apartment in Arlington, Virginia. Jerome was head of long-range weather forecasting in Washington, and he and dad spent hours talking about weather-related stuff whenever we got together. They came up and stayed in their small wooden cabin about twenty yards from ours. Lalla and Issac had built three small cabins for their three daughters so they had a place to stay whenever they visited. They were idyllic summers from 1947–59.

We spent the summers of 1960–62 in Woods Hole while dad studied mathematics courses, and I interned at the Marine Biological Labs and took science classes

Figure 4.15
Harry Wexler at Eel Pond, Woods Hole, Massachusetts, summer 1962. *Source:* Wexler Family Papers.

[figure 4.15]. Our family rented houses around Eel Pond and had a wonderful time at the beaches. Nobska beach was dad's favorite. I remember him and Jerome standing waist high in the sea at the MBL beach talking and laughing.

I guess it was his passion, enthusiasm and love of meteorology that I remember the most. The downside would be his temper, which could flare up, but not often, and his long working hours and time away on trips that made me miss him. He was a terrific dad and a great role model for me . . . optimistic, curious. I remember him driving me to the Georgetown library when we were going home from a dance lesson. I told the librarian I was picking up some books for Harry Wexler, and she set a load of about five in front of me. I looked at the top one puzzled as it was about dinosaurs. "This one must be a mistake," I said. "Oh no," the librarian said, "He wants to read about dinosaurs."

His interests were wide, his friends many and his love of life made him a magnet to his many friends and colleagues. Flying a kite in his shorts in the hot hay fields of Sharon, bundled up against the cold at McMurdo Sound, taking a break with a glass of lemonade from mowing the lawn in Falls Church are a few of my memories of dad. He was larger than life and lived life with great zest and enthusiasm.[146]

Harry Wexler indeed lived large and died young. President and Mrs. Kennedy sent a sympathy letter, as did Reichelderfer and a host of his colleagues. All the major newspapers noted his passing, and the geophysical journals eulogized him. *Monthly Weather Review* dedicated two special issues to his legacy. His career spanned the period from the launch of the first radiosondes to the launch of *TIROS 5*, its remotely transmitted cloud pictures shared with weather services around the world. He worked diligently to bring Bergen school methods of weather analysis and forecasting into the Weather Bureau and to bring Rossby's new methods into the war effort. In the postwar era, he occupied the catbird seat as a research insider centrally involved in nuclear tracers, pollution studies, weather radar, rockets, digital computing, and space observations. Those who knew him found him warm and considerate. His vision was broad, his thinking clear, and his insights well received. He was in no way singlehandedly responsible for the profound transformation of atmospheric research that occurred in this era, but his well-documented career provides a unique window into the emergence of the field. His influence lives on in the interdisciplinary field of atmospheric science.

5 Atmospheric Science

The task of the atmospheric scientists is to make quantitative measurements of the properties that *describe* the atmosphere's successive states, to *understand* the physical processes by which the successive states are determined, to *predict* future states, and, if feasible, to *influence* future states in a beneficial manner.

—Sverre Petterssen[1]

The year 1957 marked the beginning of a rapid transition from meteorology to atmospheric science and from charismatic leadership to committee work, corporate organization, and state sponsorship. It was the year the final volume of *Dynamic Meteorology and Weather Forecasting* was published, six years after Vilhelm Bjerknes's death and forty-four years since he first promised it. It was the year Carl-Gustaf Rossby passed away, suddenly and unexpectedly, creating a vacuum in inspirational leadership. Irving Langmuir also passed away that year, ending a contentious era of weather-control controversies. It was the start of the International Geophysical Year and the year that Harry Wexler invested heavily in its Antarctic program. It was also the year a National Academy committee proposed a National Institute for Atmospheric Research. In 1957, *Sputnik 1* and *Sputnik 2* orbited the planet inaugurating the new satellite age. Dwight D. Eisenhower appointed James R. Killian as special assistant to the president for science and technology and formed the President's Science Advisory Committee. A new era of expanded federal funding for science and technology soon followed.

In the early 1950s, meteorologists registered notable successes in numerical weather prediction, general circulation modeling, high-altitude soundings, and chemical climatology. In doing this, they benefitted from new technical capabilities and found inspiration in the expanded collaborative possibilities exemplified by the upcoming International Geophysical Year.

They were increasingly involved—in some cases, embroiled—in local and global environmental issues such as air pollution, weather control, and nuclear fallout and were increasingly apprehensive about the politicization of their science. In such a changing landscape, the U.S. National Academy of Sciences initiated a Weather Bureau Advisory Committee on Meteorology that soon became an influential Committee on Atmospheric Sciences. Based on its recommendations, the National Science Foundation provided funding for the University Corporation for Atmospheric Research and the National Center for Atmospheric Research.[2] To get all the acronyms out of the way at once, the IGY model inspired the NAS CAS and the NSF to support UCAR and NCAR, resulting in a vast expansion in university programs, government funding, and interdisciplinary diversity. This was only the domestic agenda. Wexler nurtured the World Weather Watch, an international program in weather service, and Jule Charney and Bert Bolin looked to the future of international cooperative research.

Atmospheric science is an umbrella term for interdisciplinary research on the composition, structure, and dynamics of Earth's atmosphere. It emerged, administratively and topically, from the union of entrepreneurial meteorologists, physicists, chemists, and space scientists seeking new sources and scales of funding for atmospheric research and better access to new tools and technologies. The new interdisciplinary field was shaped by the maturation of theoretical models, observational networks, and the needs of the Cold War. The goal was the union of meteorology, aerology, and climatology with other approaches. The scale of atmospheric science is so huge, however, that most scientists who claim professional identification with the field have only a passing familiarity with the plethora of approaches it subsumes. Individual scientists typically specialize in one area such as atmospheric chemistry, cloud physics, numerical weather prediction, general circulation modeling, satellite meteorology, applied climatology, paleoclimatology, or perhaps upper-atmosphere research. Oceanography, glaciology, bioclimatology, near-space science, and planetary atmospheres are considered affiliated sciences. Thus the unity of the atmospheric sciences is nominal, and the umbrella is huge; atmospheric scientists constitute a large tribe interested in atmospheres. This chapter provides a portrait of the field's nominal origins—the reassessment by committee, circa 1960, of national and international needs and opportunities in atmospheric science.

Committee on Meteorology

The Advisory Committee on Meteorology began its work on April 3, 1956, at the National Academy of Sciences. It was convened as an advisory committee for the U.S. Weather Bureau at the request of Louis S. Rothschild, Undersecretary of Commerce for Transportation. Members included Rossby, Horace Byers, John von Neumann, Carl Eckart from Scripps, Hugh Dryden from the National Advisory Committee for Aeronautics, Thomas Malone from the Travelers Weather Research Center, Lloyd Berkner of the Associated Universities, and retired physicist Paul Klopsteg.[3] At the first meeting, Weather Bureau chief Francis W. Reichelderfer reaffirmed his openness to a broad approach to counsel and advice on the advancement of meteorology and its applied services. Earlier advisory committees had recommended the adoption of air mass analysis and had suggested improvements in public relations and private practices. Reichelderfer sought the new committee's advice on delineating scientific research frontiers, recruiting the best talent, exchanging knowledge with other fields, and responding to new challenges posed by industrial pollutants and new ideas about purposeful and inadvertent weather and climate modification. Wexler then provided an outline of the frontiers of meteorological research, referring in particular to new methods of observing atmospheric properties, recent studies on possible changes in the chemical composition of the atmosphere and their effects, and the probable consequences of certain proposed large-scale measures of weather control. Joseph Smagorinsky followed up with remarks on meteorological research using the digital computer, including general circulation modeling of the natural and modified atmosphere. All three speakers touched on weather and climate control. The committee (Rossby and von Neumann were not in attendance) then brainstormed ways to invigorate both the Weather Bureau and the field. Their list included fellowships, conferences and summer institutes, a vigorous publication program, collaboration with other institutions, international guest speakers, and advanced university study for Weather Bureau personnel. The committee was noncommittal about the establishment of national meteorological institutes and put this agenda item on the back burner.[4]

On September 20, 1956, at the second meeting of the committee, Rossby suggested enlarging the definition of *meteorology* to include the role of the atmosphere as a carrier of pollutants and as a milieu for biological processes. He called for scientists from other fields to infiltrate meteorology to work on specific problems and to take advantage of new techniques, new challenges, and new opportunities at hand and on the immediate horizon. Detlev Bronk, president of the National Academy of Sciences, had already taken steps in this direction by adding atomic scientist Edward Teller to the committee. Soon membership was expanded to include mathematicians, chemists, physicists, and space scientists. Klopsteg suggested placing the focus on the science of meteorology and not just the immediate problems of the Weather Bureau, and in response, the word *advisory* was dropped from the committee name.

Lloyd Berkner, a consummate science policy insider, was happy to assume the chair, remarking that, "since he is not a meteorologist, his actions will be completely uninhibited."[5] His report on the recent planning meeting of the Comité Spécial de l'Année Géophysique Internationale (which planned the IGY) made a big impression. He provided details of all the upcoming projects, including plans for geophysical satellites and world-wide measurements of radioactivity. Berkner's background in physics and engineering led him to focus on the big picture: the upper atmosphere, the near-space environment, satellite instrumentation, Earth's energy balance, and climatic variations. Rossby was named cochair of the committee, with the dual appointment expected to attract scientists from other fields and from other countries into the study of meteorology. Rossby seconded Berkner's list of topics and added items to it from his recent work on atmospheric chemistry as an example of interdisciplinary expansion. He presented an ambitious plan, looking twenty-five years ahead to the next international research year in the 1980s, and asked what kind of baseline environmental measurements could and should be collected now.[6]

At its third meeting, the Committee on Meteorology discussed tropical weather systems, cloud physics, satellite instrumentation, and isotopic geochemistry. It focused on scientific education in meteorology and the ways that it could be coordinated with education in other fields of earth sciences. Berkner and Reichelderfer were at cross-purposes on this topic. Berkner emphasized knowledge and fundamental science across geophysical disciplines, while Reichelderfer wanted trained workers for the Weather Bureau.[7] The mood was somber at the March 1957 meeting at Scripps because committee member John von Neumann had died just one month earlier. This left a gap in expertise, soon filled by the appointment of Jule Charney.[8] The projected agenda for the rest of the year included a September meeting at MIT and Woods Hole to discuss theoretical meteorology and oceanography and the limits of predictability and to conduct a series of special consultations with representatives of the armed forces, the Weather Bureau, and the American Meteorological Society.

Horizons of Meteorology

On May 1, 1957, at the National Press Club, Lloyd Berkner spoke on the "Horizons of Meteorology" to a plenary session of the American Meteorological Society. Although he chaired the Academy's Committee on Meteorology, he introduced himself as a "visitor" to their ranks, offering them a view of the challenges and opportunities in meteorology as they appeared to an outsider. Meteorology was not yet atmospheric science, the International Geophysical Year had not yet started, Carl-Gustaf Rossby was still alive, and artificial satellites were not yet in orbit. Berkner had a track record of getting order-of-magnitude funding increases for research and new observing systems in ionospheric physics, polar science, and seismology. According to his colleague, the geophysicist Merle Tuve, "The astonishing thing about [Berkner's] lifetime of varied activities is the frequency with which his large-scale views and proposals were accepted and worked out, to the mutual benefits of his colleagues and the public which supported them, usually with public funds."[9]

Berkner's wide-ranging address touched on the plight of meteorologists who wrestled with difficult and abstruse atmospheric issues while facing criticism and often ridicule from the general public. He paid homage to their skills in both science and human relations. He reviewed the accelerating pace of discovery since the beginnings of systematic weather observation, a little over one hundred years prior, but focused on the "prodigious and complex" tasks that lay ahead. Scientific research is "the cutting edge of ignorance" in the sense that its practitioners are trying to understand the unknown. Berkner aimed his speech at these intellectual horizons of knowledge. He invoked the "impossible task" of predicting the behavior of the unstable atmospheric heat engine, whose variable inputs and outputs and turbulent boundaries are not well observed, especially over the oceans and the poles. He felt sorry for his meteorological colleagues who combine "rather primitive hydrodynamic equations" and their subjective judgments to predict, for no more than three days in any useful sense, the future state of the weather.[10]

In the near future, Berkner saw opportunities to improve both "comprehension and control" of the weather. First on his list of breakthroughs was the digital computer, "a new breed of fast and accurate calculating machine useful for hydrodynamic theorists and for assimilating synoptic data." Next were rotating-tank experiments in which the wave patterns vary with the rotation speed and the temperature gradients, allowing glimpses into global oceanic and atmospheric circulation patterns. Third, Berkner thought that the Antarctic program of the upcoming International Geophysical Year promised understandings at least as important as those derived from north polar meteorology of the past seventy-five years. Next, although cloud physics had made important gains since the late 1930s in understanding the formation of rain, weather control was still controversial with unsubstantiated and rather "fuzzy" claims. He thought that conserving water—"our most valuable mineral"—by suppressing evaporation and collecting melt water held the most promise. Berkner eagerly anticipated the launch of an earth satellite, scheduled for the following year, calling it "a powerful new tool that may do more to revolutionize meteorology than anything that has happened in the last century." In the planning stages were measurements of the heat balance of Earth and its variation, mapping of cloud and storm systems and their movements, and study of Earth's near-space environment. He then pointed to the practical benefits such satellites could provide in service to aviation, communication, agriculture, and shipping.[11]

Glaciologists were about to drill core samples on the ice sheets of Antarctica to document the history and current changes of climate; oceanographers were becoming more dynamic in their use of computer modeling while developing new ways of studying the ocean bottom and its sediments; and atmospheric chemists were embarking on urgent new regional and global-scale studies such as the carbon dioxide problem. Even before the start of the current measurement series by David Keeling, the issue seemed impossible to ignore: "There is no longer much doubt that carbon dioxide content of the atmosphere is increasing due to industrial burning of fossil fuels."[12] Other frontiers on Berkner's rather long list included the study of the ionosphere using radio waves and meteor trails, the study of the mesosphere using instrumented sounding rockets, and the study of the general circulation of the atmosphere and oceans using natural and artificial radioactivity.

Berkner believed that professional geoscientists as well as society at large would benefit from closer ties between practice and basic research. He supported separate international organizations for each, citing as examples the World Meteorological Organization, with member states focused on professional practice, and the Association of Meteorology of the International Union of Geodesy and Geophysics, where the intent was to pursue scientific research free of political influence. Summing up, the meteorological horizon was glowing brightly and the dawn was near, as close as July 1, 1957, when the IGY was scheduled to get underway.[13] Berkner expressed his puzzlement at the paucity of support for meteorological research, given that the promised benefits to humanity from a modest investment were seemingly great for agriculture, industry, commerce, the government, and eventually and explicitly (his laugh line), the Bureau of Internal Revenue. This was a one-off speech, but Berkner was certainly not a meteorological outsider; he was shepherding a process through the Committee on Meteorology and related institutions that would more than double federal support for the field and he had just become president of the International Council of Scientific Unions.[14]

Toward a National Institute for Atmospheric Research

Willard Bascomb, the secretary of the Committee on Meteorology, traveled to Sweden in 1957 seeking models of effective organization and interdisciplinary research. He described the International Meteorological Institute (IMI) of the University of Stockholm, popularly known as the Rossby Institute, as "not quite like any other in the world—and it is generally admired for its international flavor." The permanent faculty was small, and because the visiting members of the institute usually remained only two or three years, the composition of the group and consequently its scientific tastes and interests were always being modified or refreshed by the exchange. Bascom noted that the influence of the director, Rossby, was much felt, even when he was traveling: "Conversation would lead one to believe that he was in the next room rather than several thousand miles away at Woods Hole or Chicago." Rossby's wishes were carried out—and his administrative chores performed—by the acting director, the young and able Bert Bolin.[15]

The IMI had four divisions—theoretical meteorology, atmospheric chemistry, cloud physics, and oceanography—but the separation was not sharply made, and there was a considerable interchange of ideas in seminars and at coffee time. Emphasis, if judged by the number of people involved, was on theoretical meteorology, with a group on numerical analysis and forecasting headed by Bolin and Bo Döös. Atmospheric chemist Erik Eriksson analyzed rainwater in his laboratory and supervised a network of observers focused on acid rain and the distribution of carbon dioxide in the atmosphere and ocean. Those in the cloud physics division focused on observation and experiment. They studied convection currents near the ground, the growth of droplets, and the development of clouds and precipitation. Pierre Welander headed the oceanography division, which was developing computer models of storm tides and deep-sea circulation. The University of Stockholm paid the "housekeeping" costs and the salaries of the permanent staff, supplemented by one million kroner ($200,000) in grants from a variety of sources, including the United Nations Educational, Scientific and Cultural Organization, the U.S. Office of Naval Research via Woods Hole Oceanographic Institution, the U.S. Air Force Cambridge Research Center, the Wallenberg estate, the Swedish National Research Laboratory, and the Swedish weather service.[16]

Berkner too traveled to Sweden in late summer 1957 to visit Rossby. There the two men outlined a report for the committee meeting in September at MIT/Woods Hole.[17] They produced a document that included plans for a National Institute for Atmospheric Research. It was Rossby's final act. He died less than a week later. His institute in Stockholm provided a model for the U.S. effort but on a smaller scale. The Committee on Meteorology's admiration for the IMI was tempered by its clear understanding that they were building something quite different, less international, and not at all driven by the leader's charisma.

The committee agreed that the U.S. university system had not been able to take full advantage of new research opportunities presented by recent technological revolutions in meteorology. Roscoe Braham recalled his dissatisfaction with missed opportunities and the slow progress of research in the late 1950s. He was involved in the 1947 Thunderstorm Project, a joint government-university effort in cloud physics research based at the University of Chicago. The air force, which had supplied aircraft and radar for the project, could no longer provide such large-scale support. Also, big computers were rare and exceedingly expensive. Braham felt that if the universities had access to greater pooled resources, they could do a much better job of basic research on the atmosphere.[18]

In November 1957 and again in January 1958, the Committee on Meteorology recommended the creation of a national institute and an order of magnitude increase in support for meteorological research. Wexler wrote in his Antarctic Diary for January 17, 1958, "Just returned from two days in New York and Brookhaven with the National Academy of Science Committee on Meteorology to discuss more or less final details for the plan to [establish] a National Institute for Atmospheric Research."[19] On February 5, 1958, four months after the Soviet Union launched its first Earth satellite and five days after America launched *Explorer 1*, the committee formally presented its first report on "Research and Education in Meteorology" to the Academy with copies to the executive branch, including the President's Science Advisory Committee, the National Science Foundation, and the departments of defense and commerce. The report waxed poetic about how the atmosphere is both our life-support system and a natural "laboratory" for studying electrical, chemical, optical, thermal, acoustical, and mechanical phenomena. It asserted that meteorology and related sciences were poised for new breakthroughs, that they had been seriously neglected, and that the national interest demanded rapid and vigorous investments in research education and personnel. At the time, there were seven universities in the country with meteorology departments that offered a PhD and four others where a PhD in meteorology could be earned in other departments.[20] Ninety percent of the nation's 5,273 weather professionals, most without the PhD, were working in civilian and military branches of the government; colleges and universities employed only 135. The report recommended an immediate increase in support for basic research at universities of 50 to 100 percent. It also formally recommended the establishment of a National Institute of Atmospheric Research that would be "independently operated by a corporation sponsored by a group of universities under a prime contract with the National Science Foundation." The report suggested funding of $50 million for the first five years.[21]

The committee was not alone in its advocacy. In 1958, the American Meteorological Society began a comprehensive program to attract more qualified students into the discipline. That same year, the National Science Foundation established an Atmospheric Sciences Program aimed at providing enhanced, focused, and stable funding in the field.[22] The program was informed by successes of the IGY and was necessitated by NSF's new responsibility for coordinating federal weather modification programs. The discussions were also motivated by fear. There was a general sense of angst that the United States was somehow involved in a "weather race" with the Soviet Union that might lead to weather control or even weather warfare. The pundits, the press, and many cold warriors had a field day with this, but even mainstream scientists like Henry Houghton at MIT expressed concerns that the Russians might someday be able to attempt an unfavorable modification of the U.S. climate in the guise of a peaceful effort to improve their own.[23]

In February 1958, Thomas Malone convened an ad-hoc meeting of scientists from major institutions involved in meteorological research to discuss and then endorse the committee's report, especially the idea of a national institute. Columbus Iselin of Woods Hole was the only participant directly opposed to the institute idea, but others had concerns that staffing the institute might be a problem. There was general agreement to emphasize humanitarian rather than military-related aspects of meteorology. Moving forward with intellectual support and initial funding from the National Academies, Houghton convened a new University Corporation on Atmospheric Research (UCAR) with representatives from fourteen universities with meteorology programs—Arizona, Chicago, Cornell, Florida State, Johns Hopkins, Michigan, MIT, NYU, Penn State, St. Louis, Texas A&M, UCLA, Washington, and Wisconsin. During the annual awards dinner of the New York Board of Trade at the Waldorf Astoria in October, Houghton, speaking on behalf of the UCAR-associated universities, publically announced plans to form a National Institute of Atmospheric Research and to seek funding from NSF.[24]

In the summer of 1958, the Committee on Meteorology voted to change its name to Committee on Atmospheric Sciences. The vote was not unanimous. Malone thought that, given the success of the progress report recently issued under the current committee name, a name change might confuse government officials just as new opportunities for meteorology were opening up. Also, the original charge of the committee was "to vitalize this science and foster closer relationships with other areas of science on which the future of *meteorology* may depend." Malone welcomed the participation of scientists from other disciplines but worried that by doing so the traditional practices of meteorology might be somehow eclipsed. Finding no opposition from the Weather Bureau, which funded the committee, the name change was approved.[25]

Bronk addressed a large Scientific Information Meeting on Atmospheric Sciences in November 1958, expressing his appreciation for the interdisciplinary, international, and intergenerational work being undertaken. He noted that their deliberations brought together many fields of science, dealt with a subject bigger than any one country alone, and by spinning off UCAR, provided a bridge between training and research at the universities and the needs of government. He borrowed Baconian analogies from "Solomon's House," referring to the proposed National Institute as a cooperative intellectual scientific "monastery," and alluding to its scientific "merchants of light" who would go forth to exchange ideas and develop plans. The following spring, UCAR formally became a corporation with Houghton as chair, Byers as vice chair, and M. A. Farrell of Penn State as the secretary-treasurer. They began planning the steps needed to build a National Center for Atmospheric Research (NCAR). Byers and Malone served on both the Academy committee and UCAR; Byers was a Rossby protégé; all three men were MIT graduates.[26]

Regular meetings of UCAR resulted in a comprehensive report, colloquially called the NCAR Blue Book, outlining the next steps to be taken.[27] These included studies of (1) atmospheric motion, encompassing the entire spectrum of wind systems from the general circulation to turbulent eddies, tropical and extratropical cyclones, hurricanes, tornadoes, squall lines, and anticyclonic wind systems; (2) energy exchanges, through radiation, horizontal and vertical mass transport, and the exchanges of sensible and latent heat; (3) water in its liquid, solid, and gaseous phases involving chemical and physical processes in clouds, ice sheets, bodies of water, and precipitation; and (4) physical processes involving electrical, chemical, optical, and acoustical phenomena in the lower and upper atmosphere. The National Institute was to be equipped with a flight facility, computing center, library, radiation and remote sensing facilities, shop, and a suitable building. Within five years, UCAR anticipated an annual budget of $13 million and a staff of 550, including 108 scientists.

UCAR further recommended that the national center be built on a scale sufficiently large to investigate the global nature of fundamental atmospheric problems, provide facilities, technological assistance, and interdisciplinary scientific talent beyond those possible by individual universities, and preserve the alliance of research and education. Overall, the UCAR staff held seventeen topical planning conferences all over the country to review the state of knowledge in particular fields and envision how a national center might support and enhance research.[28] UCAR formally became a corporation composed of its member universities on March 16, 1959, in Delaware. Twenty-five representatives from fourteen university members gathered at the University of Arizona, Tucson, for their first official meeting on April 2, 1959. Meteorologist John Dutton recalled a quarter century later, "The founders of UCAR created the new organization as a grand experiment based on a simple hypothesis: university atmospheric science could be more effective through collaborative efforts."[29]

"Good Enough"

In the summer of 1959, UCAR sought a director for the new National Center for Atmospheric Research, and James Van Allen was on the top of the list. Two years earlier, he had turned down an invitation to join the Committee on Meteorology. Van Allen, forty-five years old at the time, was a student of terrestrial magnetism and space physics, an instrument builder who designed radio-proximity fuses for the navy in World War II, a flier of instrumented sounding rockets and balloon-borne rockoons, and most recently, the chief scientist for the *Explorer* satellite series that had discovered Earth's magnetosphere. The "Van Allen belts" were named for him, and he had just appeared on the cover of *Time* magazine.[30] He was comfortably ensconced at the State University of Iowa and identified himself as a space scientist looking to the solar system rather than an atmospheric scientist looking at clouds. He held out some interest in the NCAR directorship but ultimately declined. The search continued with a new short list that included Walter Orr Roberts, Michael Ference Jr., Herbert Friedman, and Thomas Malone, the latter being the only candidate with formal training in meteorology.[31] Roberts was a Harvard-trained astronomer, director of the Harvard-Colorado High Altitude Observatory (HAO), with research interests in sun-Earth interactions; Ference, a physicist and director of the scientific laboratory of the Ford Motor Company; and Friedman, a physicist with the Naval Research Laboratory and pioneer in the use of sounding rockets for aeronomy and astronomy. Wexler, whose forté was meteorology and research management, apparently took himself out of consideration for the job.

After a year of deliberations, on June 27, 1960, Alan Waterman, director of the National Science Foundation, and Henry Houghton, chair of UCAR, announced the appointment of Walter Orr Roberts, age forty-four, as first director of NCAR. Waterman also announced the awarding of a $500,000 contract from NSF to begin planning for the facility. With this appointment and the new grant, atmospheric science had turned a corner. Many viewed Roberts as an able scientist, scientific administrator, and science fundraiser with strong interdisciplinary interests across the atmospheric and space sciences. These were precisely the talents most valued in the age of committees. He reached prominence in Washington-based science policy and decision-making circles, serving on the UCAR board and then the NSF Atmospheric Science Board, making decisions about an institute he would later lead. He had no background in weather studies but was well positioned to join the quest for government funding for research.

With four or five alternative locations for NCAR in play, Roberts made a concerted push to see that his home town would be the chosen site, sweetening the deal with financial support from his university, the city of Boulder, the state of Colorado, and a choice piece of real estate on Table Mesa along the front range of the Rocky Mountains. Richard Kassander Jr., who served on the search committee, recalled Roberts's political savvy, insisting, as a seemingly reluctant (or perhaps coy) candidate, that the High Altitude Observatory would have to become part of NCAR and NCAR would have to be located in Boulder, knocking competing locations out of the running. When asked by the search committee what he would do if NCAR did not wind up in Boulder or HAO did not eventually become part of NCAR, he replied that "you could find yourself another director."[32] In 1960, the National Science Board approved the site and an expanded operating budget.

Roberts turned out to be a popular choice. He was known for his openness, generosity, enthusiasm, and his breadth of vision and inclusiveness. He was gregarious and well liked, not least for the Sunday evening musicales held at his home. In his obituary, he was called "an international science statesman" whose talent lay in creating organizations.[33] Yet throughout his life, Roberts held deep insecurities about his scientific abilities; he was engaged in a constant quest for funding; and he adopted a managerial approach to research, even in the early days when he was managing just himself as the sole researcher at the HAO. Roberts eventually imposed his organizational vision for NCAR on the building design, with its soaring isolated towers reserved for "researchers pursuing individual projects of their own choice." Working closely with architect I. M. Pei, he became, in effect, "an architect of atmospheric science," at least of early atmospheric science.[34] A broader sense of atmospheric science, however, would have to come from other sources, including the numerous UCAR universities.

A quarter century later, Jule Charney was asked in an interview if he thought it was a good idea to establish NCAR. He basically replied that it turned out to be "good enough" but that he had "great qualms" in the beginning. He had joined the Committee on Meteorology in the summer of 1957 and served under the leadership of Lloyd Berkner and Paul Klopsteg. He credited Klopsteg with broaching the subject of a separate government-supported meteorological research institution and, in support of this idea, he had personally interviewed such influential scientists as Columbus Iselin, Edward Teller, and Harold Urey to gauge their reactions. Charney recalled his former supervisor at the Institute for Advanced Study, J. Robert Oppenheimer, saying that these things succeed when they are done in good style and that the way in which these institutions are inaugurated will influence their subsequent development. Charney was of the opinion that a national institute should establish a standard of excellence and that the universities should be willing to sacrifice some of their members to establish a strong center that would work in cooperation with the universities and not completely dominate them. As Charney remembered,

That was a touchy problem. I could not at that time accept the notion that such an institute would be justified as a garage for aircraft or even [a building] for a large computer. Since then I've changed my mind somewhat. In other words, I think that the fact that an institution of this kind could make facilities available that would not be available to a university group has become an important raison d'être. I think that where the institute has probably fallen down—and maybe it was not possible in an institute run by the universities—[is] to have an establishment that was clearly superior intellectually. But I think it has been good enough intellectually to attract people from the outside, university people, as a center. . . . I think it has been essential that they have in-house people who are doing interesting work to attract students, but also to help in meetings, and joint enterprises with the universities. So, on the whole—I've always been very ambivalent about such things and expressed it in the early days—but in the end I think that it has served its purpose.[35]

Charney had very little contact with NCAR over the years and never used its facilities. He would have preferred if it had become more of an institute for advanced study rather than a center of support for other people's projects.[36]

New International Horizons

The frenzy of committee work regarding the institutionalization of the atmospheric sciences occurred in the second term of the Dwight D. Eisenhower administration, with the active oversight and involvement of a large number of agencies and the invention of a number of new ones. Projects to build and expand national efforts in scientific research and to coordinate federal efforts were foremost on the minds of U.S. leaders in the post-*Sputnik* era. Building a world weather reporting and prediction system was near the top of the list of desired projects. It involved enhancing existing national weather services using satellite observations and global interconnections via teletype. Next was a research program of global scope to follow the International Geophysical Year. Plans included satellite measurements of Earth's heat balance, focused studies of storm dynamics and predictability, and a world climate survey to monitor climatic changes and quantify land and water resources. Education and training were also targeted for dramatic expansion in the United States and worldwide. UCAR and the WMO were to be at the center of it all. Such a coordinated U.S. national and international effort under the banner of the atmospheric sciences was intended to combine service, research, and education. Cooperation was needed in a shrinking, more dangerous, yet more interdependent world; these were worthy projects for both scientific and political realists and idealists.

When John F. Kennedy took office in January 1961, his science adviser Jerome Wiesner asked the Academy to conduct a study of research opportunities in the atmospheric sciences. They, in turn, asked Sverre Petterssen, a product of the Bergen school, to lead a decadal study. In his long and distinguished career, Petterssen had worked closely with Vilhelm Bjerknes, Carl-Gustaf Rossby, and Harry Wexler. He had been an apprentice of Tor Bergeron, director of the Bergen forecasting center, chair of the meteorology department at MIT, chief forecaster for British aviation during the war effort in Europe, head of the Norwegian Forecasting Service, chief meteorologist for the International Civil Aviation Organization, director of scientific services for the U.S. Air Force Weather Service, and, at the University of Chicago, professor and chair of the department of meteorology and first chair of the new department of geophysical sciences.[37]

The final report produced by Petterssen's committee, "The Atmospheric Sciences, 1961–1971," emphasized the intrinsic scientific merit of the science and the excitement of the research frontier. It focused on the domestic agenda and contained a section on international cooperation. The report called for a tripling of federal expenditures for research in atmospheric science to $600 million, a tripling of active scientific personnel, a quadrupling in the production of doctorates, and the creation of "centers of excellence" that would allow aeronomy to merge with meteorology into a unified atmospheric science program. Atmospheric science, in turn, should become an integral part of future graduate programs in geophysics and be closely associated with first-rate work in classical physics. The report concluded by linking the progress of the domestic agenda with effective international cooperation; it did not mention the military needs of the Cold War.[38]

The domestic agenda, however, lacked clarity. The federal agencies, each with vested and expanding interests in the atmosphere, remained uncoordinated and unorganized regarding their activities in research, development, and practical applications. The relationships between weather and society had changed. Francis W. Reichelderfer took the lead in articulating this:

Until 1940 the concept of meteorology was principally one of climate or weather. The impact of weather upon agriculture was the major concern. . . . Since 1940 the concept of the atmosphere is not simply as the breeder of weather. The atmosphere has become a more intimate environment of mankind, a course for its vehicles and a potentially benign or even a malignant factor to be studied and possibly influenced.[39]

Reichelderfer reinforced these sentiments in 1963 in his official letter of retirement to President Kennedy after a quarter century at the head of the Weather Bureau:

This span of years has brought many changes in knowledge of the science of the atmosphere and in the technologies for applying it to human desires. It has been my good fortune to be here during the exciting period of growth of applied meteorology almost from its infancy as an earthbound, manual operation to its recently-acquired capabilities with spacecraft like satellite TIROS and with the most advanced, high-speed computers to digest the worldwide data and open heretofore undreamed future capabilities for beneficial uses of knowledge of the atmosphere and its weather and climate.[40]

A domestic strategy in the atmospheric sciences was clearly needed so that federal agencies could realign their missions with the times and coordinate efforts. Two steering committees worked in parallel with the Petterssen committee—the Panel on the Atmospheric Sciences within the Office of Science and Technology, chaired by statistician John W. Tukey, and the Interdepartmental Committee on Atmospheric Sciences of the Federal Council of Science and Technology, coordinated by the NSF and representing a wide range of agencies spanning the federal government.[41] Petterssen also convened an Ad-Hoc Committee on International Cooperation in the Atmospheric Sciences, whose members included Wexler and Charney. The committee recommended constant surveillance of the entire atmosphere:

The atmosphere represents an international resource of fundamental importance to human activities . . . [that] must be studied on the global scale and at all heights above the earth. Progress in the atmospheric sciences depends not only upon advances in the physical sciences in general but also upon technological developments, and both areas have gone forward with extraordinary speed in recent years . . . making it possible, for the first time in history, to keep the whole atmosphere under constant surveillance. [The time is right] for a concerted effort to observe, describe, and understand the atmospheric processes and, thus, to place weather forecasting on a firmer scientific basis and to consider the possibility of modifying weather systems on a large scale.[42]

The committee emphasized humanitarian and practical considerations, including improving weather observation and prediction to protect lives and property, providing fresh water for a growing world population, addressing problems of air pollution (including nuclear fallout), and collaborating on studies of large-scale weather control. Weather services had long been important for agricultural production, mainly on local, regional, or national levels. Now the needs of a growing world population for food, water, and effective land management were coming into play driving a demand for improved understanding and even control of climate on a global scale. The new weather satellites orbiting the globe passed over hundreds of national borders each day, symbolizing the need for national weather services to enter a new international era of data collection and resource sharing. In a world recovering from world war and experiencing

Cold War tensions, this would require new national institutions and new international agreements.

The Petterssen committee recommended establishing a truly global observing system in support of both meteorological services and atmospheric science. This would involve an improved and expanded network of weather observing facilities along with international standards for data processing and communication systems. Meteorological satellites would provide increased coverage over sparsely populated areas (such as oceans, polar regions, and deserts), and a dense network of radiosondes and rocketsondes would extend coverage to high altitudes. The data obtained would be made available to the weather services and research scientists of all countries. These ideas informed President Kennedy's proposal to the UN General Assembly, the United Nation's subsequent passage of Resolution 1721 on international cooperation in meteorology and the peaceful uses of outer space, the World Weather Watch, and the designation of the WMO to take the lead in both weather research and service.[43]

In April 1962, Charney circulated a two-page letter to Wexler and a number of colleagues responding to the recent UN resolution calling for an international research program to "advance the state of atmospheric science and technology so as to provide greater knowledge of basic physical forces affecting climate and the possibility of large-scale weather modification." He then announced a one-day informal meeting sponsored by the Committee on Atmospheric Sciences to help shape the program, one he projected would entail an expensive and extensive effort involving a large number of scientists. His letter included a substantial proposal outlining "A Suggested Meteorological Observation Plan" for measuring the *entire atmosphere* synoptically. Charney contemplated a limited period of intense coordinated observations, "long enough to furnish a scientifically and operationally significant pilot study, but short enough to remain economically feasible." His three-part program involved building up the instruments and the capabilities for a global observing program over a five- to ten-year period, an intensive measurement program for up to a month, followed by data processing, modeling, and machine computation. He wanted to avoid "the fragmentary observations taken in the International Years or in regional observational studies," where the data languished in archives. Charney drew the analogy to the limitations of measuring only a small part of the fluid motions in a rotating heated tank rather than taking comprehensive measurements including the heat flow through the boundaries and implied that better understanding and longer-range forecasts would result from complete Earth system measurements. Global measurements would allow new experiments, including treating the Southern and Northern Hemispheres as separate experimental regions with different boundary conditions: "Here is a large-scale climate modification experiment that nature places in our hands." Global data would also reveal interactions among atmospheric layers, between ocean and atmosphere, and between the two hemispheres. It would allow a greater focus on the tropics where the transfer of heat, momentum, and kinetic energy originates, specifically in tropical currents and hurricanes.[44]

Given the nature of atmospheric science, where prediction and the possibility of control are correlatives of understanding, Charney expected practical benefits to be forthcoming, including improved short- and long-range prediction and understanding of climate, improved instruments, international standardization, intercomparison of forecasting techniques, regional as well as global data for area studies, stimulation of meteorological education and training, and the possibility of evaluating weather and climate control under international auspices. He also envisioned a parallel program for global ocean measurements. Charney was thinking creatively about what would become the Global Atmospheric Research Program, in parallel with Wexler's investment in the World Weather Watch.[45]

Soaring Aspirations, Crushing Limitations

Atmospheric science emerged after 1957 and made great strides in only a few years. Many of the key leaders of the committees that deliberated and formulated the institutions of atmospheric science were not meteorologists, and those that were gravitated more toward administration than research. Lloyd Berkner, Paul Klopsteg, Henry Houghton, Horace Byers, Alan Waterman, Walter Orr Roberts, Thomas Malone, and Sverre Petterssen worked incredibly hard on administrative structures, support, and even nitty-gritty details. But with the exception of Roberts in his own folksy way and Berkner in the centers of power, they were not particularly charismatic or visionary, and they certainly did not represent the cutting edge of research. Jule Charney and Harry Wexler did. Their thinking benefitted from all the input from the committees, but they also played important roles in shaping this input in important ways. Decades later, the intense period of activity between 1957 and 1962 that gave rise to atmospheric science resulted in a viable UCAR, a venerable NCAR, a fully implemented World Weather Watch, the innovative Global Atmospheric Research Program, and new warnings about climatic change and the limits of predictability. This is not the end of the story but, in essence, a new beginning because meteorologists aspiring to portray "everything atmospheric, everywhere, always" and to issue sufficiently accurate weather forecasts had (and still have) to come to grips with the radically new conceptual world of chaos theory.

At a landmark conference on numerical weather prediction in Tokyo in November 1960, Edward N. Lorenz presented a landmark paper that would set the agenda for the next decade and beyond. Charney was there, as were Jerome Namias, Hsiao-Lan Kuo, Arnt Eliassen, Ragnar Fjørtoft, Bert Bolin, Norman Phillips, George Platzman, Joseph Smagorinsky, and 125 other participants, fifty of them from abroad. Lorenz, a 1938 graduate in mathematics from Dartmouth College with a 1940 master's degree in mathematics from Harvard University, served as a meteorology instructor at MIT during World War II, earning his ScD there under Victor Starr in 1948. Lorenz knew Rossby well and had an office in the department of meteorology at MIT near Charney and Phillips. His research for the Statistical Forecasting Project involved running a simplified two-layer baroclinic model to generate a series of computer generated "weather maps." He then attempted to repredict these maps using linear regression methods, and the results were a game changer. His technique produced excellent forecasts one day in advance, mediocre forecasts more than three days in advance, and no results at all for long-term forecasts. During the discussion of his paper at the Tokyo meeting, Bert Bolin asked him, "Did you change the initial condition just slightly and see how much different results were in the forecasting in this way?" Lorenz responded that he had changed one of the twelve variables by a factor of a small fraction of 1 percent, an amount considered to be less than observational error. He found that this error grew and continued to grow at a slow exponential rate until there was no resemblance at all between the initial and final maps. This implied that, at least for this particular set of equations, there is a forecasting limit. He had identified the principle that weather systems have a "sensitive dependence on initial conditions," which is the founding insight of chaos theory. Lorenz's work on the unpredictable behavior of complex natural phenomena propagated through and far beyond meteorology, inspiring a wave of interdisciplinary investigation into a new universe of nonlinear systems and leading to new understandings of a broad range of phenomena—not as disordered, discontinuous, and erratic but as ultimately beautiful and clearly governed by a new set of concepts. Any history of the development of atmospheric research after 1960 will have to include the fundamental contributions of Ed Lorenz.[46]

6 Final Thoughts

In the case of Science there is no other mirror than History. . . .
—Tor Bergeron, 1959[1]

During the first six decades of the twentieth century, three generations of scientists—personified by Vilhelm Bjerknes, Carl-Gustaf Rossby, Harry Wexler, and their close associates—invested their life energies in advancing the cutting edge of atmospheric research. They transformed meteorology into a physical science, built new theoretical models, established new institutions, trained new people, and incorporated the most promising new technologies into the atmospheric sciences. Their lives span a full century, their work spans a period of technological flux, from Marconi wireless and the Wright Flier to digital computing and weather satellites and from roentgen and Becquerel rays to outdoor nuclear testing. They aspired to build ever larger, even global, networks of observation and organization; to attain near perfect measurement, near perfect understanding, near perfect prevision; and to make their science as rational and useful as possible. Their life stories and accomplishments are interwoven into a sixty-year big picture history of atmospheric research, a tapestry formed from the braided threads of technology, science, and social change.

The Gordian knot of meteorology—that intertwined tangle of imprecision and nonlinear influences that, if unraveled, would provide prevision of the weather for ten days, seasonal conditions for next year, and climatic conditions for a decade, a century, a millennium, or longer—was never untied, even if it was sometimes "cut" by new technologies, to use Rossby's terminology. Instead, the work of Edward N. Lorenz showed that the "knot" analogy was much more subtle than previously thought and that if it even existed, it could never essentially be untied. Lorenz introduced the novel understanding that chaos theory applied to meteorology was a new topology, an extreme sensitivity to initial conditions in a dynamical system of deterministic nonperiodic flow.

In later years, Lorenz explained how chaotic behavior manifests itself in rather simple systems, such as a skier on a slope, as well as in exceedingly complex systems, such as in global weather and climate models containing a near infinite number of variables. He traced the conceptual roots of chaos to Henri Poincaré's work on the three-body problem, to Lewis Fry Richardson's quixotic attempt to integrate the primitive equations, to the shifting patterns produced in wave-tank experiments when minute changes are made in rotation speed or heating, and to the computational difficulties such as those encountered by Norman Phillips in his early global circulation model. Lorenz found that the weather, long thought to be predictable based on the laws of hydrodynamics, was actually a chaotic system. He also made clear that neither was it a random system. This empowered a new quest by meteorologists to discover the atmosphere's underlying dynamical laws. Lorenz's 1960 conference paper in Japan marked the beginning of the end of an era, but it was only the opening bell of a challenging quest that continues today to incorporate chaos theory into atmospheric science.[2]

Was the period from 1900 to 1960 in meteorology more exciting than the century before? Having also written about the earlier era, I think it is safe to say it was. Individuals and ad hoc groups of observers, national weather services and international associations, and the history of telegraphy dominate the story of nineteenth-century meteorology, an era in which weather service directors and national capitals—London, Berlin, Paris, Washington, DC, and Saint Petersburg—were at the center of most of the activity. For the most part, the distinguished theoreticians and practitioners of the science in the nineteenth century worked alone or through correspondence networks to expand the horizons of meteorology.

Things were different in the new century. Twentieth-century meteorologists were the beneficiaries of a number of fundamental breakthroughs in electronic communications, aerospace capabilities, and nuclear power. Their centers of excellence were in the scientific institutes of Stockholm, Leipzig, Bergen, Cambridge, Massachusetts, and Chicago; their centers of action in Washington, DC, Princeton, and Boulder. Most meteorologists knew one another, worked together more than occasionally, and were trained, mentored, supervised, and inspired by the three main protagonists—Bjerknes, Rossby, and Wexler. The leaders of atmospheric research in this era linked theory and practice as they reached for the tantalizing brass ring of progress, which was just within reach, just out of reach. Their excitement was palpable, and it was present in every decade.

The atmosphere of 1900—what meteorologists knew or could know about it and how society made use of it—was profoundly different from that of 1960. This is true, in fact, of every decade in between. Commercial and military developments in radio communication, aerial and aerospace transportation, and nuclear technologies did not just provide new tools but expanded the volume of the atmosphere dramatically and transformed the basis on which atmospheric scientists conducted their research.

Vilhelm Bjerknes made meteorology into a physical and mathematical science of fluid mechanics and thermodynamics. His focus moved from ideal to real fluids, from ether to air, and from the abstract realms of divergence and curl to the practical needs of farmers and fishers in West Norway and weather forecasters in most every nation. His vision encompassed air masses as large as continents and a polar "uberfront" girdling the planet, spinning off wave after wave, family after family, of cyclones with cold, warm, occluded, and stationary fronts. He and his many students and associates became the leaders of a recognizably modern meteorology in which intractable mathematical problems were handled by graphical, accessible, and memorable methods. His career peaked in the era of wireless radio and the first heavier-than-air flights, and it was interrupted by two world wars, but he lived to see the dawn of digital computing, scientific rocketry, and widespread commercial aviation.

Carl-Gustaf Rossby, too, moved in both geophysical and practical circles, first as a Bergen school emissary and then as a builder of models of planetary waves that circle the entire Northern Hemisphere at high levels. He inspired and mentored a new generation of theorists. During World War II, he trained a host of dedicated weather officers and consulted on practical meteorological matters worldwide. His toolbox included airplane soundings, radiosondes, digital computing, and networks for chemical climatology. He paid considerable attention as well to smaller-scale phenomena, such as turbulence and fog, both important in aircraft safety. He built three major academic programs—schools, really—in Cambridge, Massachusetts, Chicago, and Stockholm; launched several journals; and circulated in the highest levels of government. Wherever Rossby went, and he went far, he found himself involved in intense discussions on atmospheres and oceans, surrounded by his many colleagues and nurturing a cadre of elite students.

Harry Wexler was a student of Rossby and might be called a "grand student" of Bjerknes. He started out working on anticyclones and cold air masses that were as interesting, important, and ubiquitous as cyclonic storms. The laws of radiation governed their thermodynamic structure and the laws of hydrodynamics their motion. Cold air outbreaks had important social consequences, typically bringing good flying weather, triggering frost warnings and winter heating emergencies, and, when they stagnated, intensifying air pollution events. Wexler worked to bring air mass analysis into the U.S. Weather Bureau, trained thousands of weather officers in practical methods during World War II, and then occupied the "catbird's seat" as head of research in the Weather Bureau. His portfolio included the use of radiosondes, radar, high-altitude sounding rockets, nuclear tracers, digital computers, and satellites. He established the organizational foundations of operational numerical weather prediction and general circulation modeling. He headed U.S. research programs for the International Geophysical Year; established radiation, ozone, carbon dioxide, and other measurements in Antarctica and at the Mauna Loa Observatory; and developed the U.S. weather satellite program, including *TIROS* and the global heat-budget experiment on *Explorer 7*. His leadership in satellite meteorology inaugurated the long-anticipated new era of panoptic Earth observations from space. During the final year of his life, the chilliest of the Cold War, he served as the chief U.S. negotiator for the World Weather Watch, a project aimed at the peaceful use of outer space.

All three of these individuals sought balance between their theoretical interests and practical applications, and they flourished during a remarkable era of rapid technological change. In their own times and in their own ways, Bjerknes, Rossby, and Wexler inspired, nurtured, led, and willed into existence important aspects of the modern atmospheric sciences. They informed, stimulated, and in some cases led the conversations that resulted in meteorology's turn to a more interdisciplinary atmospheric science.

Big picture history is an apt vehicle for presenting the story of atmospheric research. It requires proper framing and considerable attention to both composition and to the details of the life and work of warm-blooded

protagonists as revealed in new archival sources and in the existing historical and technical literature. What has been learned, discovered, or established during these heuristic sessions at the loom of history? Here is my list.
I welcome your additions.

1. The main protagonists—Bjerknes, Rossby, and Wexler—were certainly
   not lone geniuses. They, along with their many associates, were talent
   scouts and public-minded builders of something much bigger than
   themselves who brought meteorology into a new era of science and
   practice. They were risk takers with grand visions of what the science
   needed and what it could become. With support from the Carnegie
   Institution, Bjerknes launched the careers of his numerous talented
   assistants, worked diligently to expand and improve observations, and
   when Norway needed him most, established a practical school of forecasting. He did this *in addition* to being the top theoretical meteorologist of his day. Rossby was a brilliant theorist and a talented organizer,
   founder of a new school of atmospheric research, and founder of at
   least three of the top schools of postgraduate training. When his
   adopted country needed him for the war effort, he rose to the challenge; when Sweden, Europe, and the world needed him in the postwar
   era, he answered the call as well. Wexler saw the global picture of atmospheric research emerging. He knew that new technologies were rapidly revolutionizing theory and practice in atmospheric research and
   that the future lay in the interweaving of all the new capabilities, not
   in any single one. He worked himself, literally, to death to provide the
   necessary foundations and frameworks.

2. The goal of meteorologists, as expressed in 1960, to portray "everything atmospheric, everywhere, always" was not merely a final goal but
   infused the perceptions and animated the actions of three generations
   of researchers. It is a distant echo of the *réseau mondial* proposed in
   1896 by Swedish meteorologist Hugo Hildebrandsson. It is embedded
   in Bjerknes's notion that the polar front is a global phenomenon
   extending to the top of the troposphere (chapter 2), in Rossby's
   preference for theoretical analyses of hemispheric flow patterns and
   the global circulation (chapter 3), and in Wexler's claim that international cooperation made sense because weather phenomena "affect
   everyone everywhere" (chapter 4). The radar meteorologists wanted to
   "measure everything, everywhere, all the time," as did the satellite

meteorologists, with their Earth heat-budget experiments and "truly global weather patrol." The modelers of the general circulation of the atmosphere and oceans sought to put the world or at least its governing equations into their digital machines. These technologies did not untie the Gordian knot of forecasting but at least cut major cross sections through it.[3]

3. Vilhelm Bjerknes was not a failed physicist but a classical one who made the transition from studying ideal to real fluid motion. He was a physical scientist working to provide building materials for the final edifice of the science of meteorology. He did not break free of his father's influence but returned throughout his life to the analogy between hydrodynamic and electromagnetic fields of force, repeatedly invoking the concept of ether, at least metaphorically. Bjerknes grew up thinking about an ideal atmosphere, homogenous and incompressible, mathematically elegant and governed by laws of rotation and divergence similar to those assigned to electromagnetic radiation. In midlife, he shifted his attention to the real atmosphere in all it messy baroclinicity, setting out on a quest to measure its current state and predict its future states. His strong convictions established a well-defined if distant goal for theoretical meteorology.

Bjerknes also led a systematic attack on the forecast problem—improving the collection of observations and developing rational methods to record, display, and interpret them. In later life, sobered by the immensity of the task and responding to practical needs, he founded a dominant school of practical forecasting based on disturbances propagating along a "polar front" and graphical methods to study them, developed by his son Jacob, Halvor Solberg, Tor Bergeron, and a host of others. Members of the Bergen school spread their message and methods to the weather services of the world through lectures, publications, exchanges, and in many cases emigration and settlement in the major international centers of meteorology.

With the assistance of his core team, Bjerknes produced his magnum opus, *Physikalische Hydrodynamik*, in his seventy-first year. In his eighty-eighth year, he would have received news of a successful computer forecasting experiment conducted in Aberdeen, Maryland, but then would have turned wistfully back to his equations because the forecast had been for a *barotropic* atmosphere. Bjerknes's macroquixotic quest to chain Laplace's demon (or untie the Gordian knot of forecasting: pick your metaphor) did not lead to practical success in meteorological forecasting. The reigning gold standard in forecasting—frontal and air mass analysis as practiced by the Bergen school—was, in its time, the latest word, but it was not the final word. However elegant, Bergen school methods do not contain the dynamic and thermodynamic principles needed for calculating the trajectories and evolution of weather systems for more than several days. Rossby's methods, written out in tractable form by Charney, were the first to be computerized. They were, at once, physically based (with appropriate simplifications), quantitative, and practicable. They produced rather impressionistic numerical forecasts of the flow pattern at midtroposphere level, not complete weather forecasts. Remember, the 500-millibar forecast model was vastly simplified, excluding humidity, radiation, and stability considerations—all observable factors. Although the twenty-four- to forty-eight-hour numerical forecasts were impressive, an experienced human forecaster at the time could match their accuracy.

In 1959, Tor Bergeron issued a long-range forecast for weather forecasting itself, warning his contemporaries that they might become complacent, lulled into thinking they had "arrived" at a kind of digital promised land. Weathermen in the near future might see no point in performing synoptic analyses of what he called "the weather itself" based on the composite map. He predicted they would come to rely on two *dei ex machina*—the electronic computer to do the drudge work and general dynamic models, however simplified, to do the conceptual work. According to Bergeron, the temptation was at hand for a younger generation to deem the new methods "objective" and "thoroughly modern" and to throw out the old "subjective" methods.

Bergeron continued his warning. Although correlations between 500-millibar patterns and surface weather were tenuous, meteorologists seemed rather content with the upper-air models that worked best for the most quiescent, quasi-stationary weather situations—cases in which weather forecasts are trivial. Electronic computers were not yet able to calculate great and sudden departures from the stationary state—for example, the beginning or end of fine weather conditions, extreme temperature departures, or major weather calamities. But those "critical forecasts" are the cases of greatest importance to the public.

Bergeron's biggest lessons of history included avoiding complacency with current methods or being dazzled by new ones; keeping channels of communication open among theorists, empiricists, and practitioners; and seeking (and optimizing balanced relationships among) the best observations, tools, techniques, and models. He strongly advocated using the new methods for all they could provide, without giving up or forgetting to use older methods. His hopeful forecast was that the mathematical methods employed by Rossby and Charney and the craft methods of the Bergen school might fuse into a "higher unity."

4. The pioneering work of Anne Louise Beck was newly uncovered during the course of this research. During her fellowship year in Bergen, she served as Bjerknes's editorial assistant for his most significant paper, "On the Dynamics of the Circular Vortex." Her effort to bring Bergen methods to the U.S. Weather Bureau in the early 1920s was a first step in a process that ultimately required the efforts of many workers over two decades. In Beck's *Monthly Weather Review* article, "Earth's Atmosphere as a Circular Vortex," her depiction of fronts and a cyclone on a U.S. weather map was completely distorted by the heavy-handed manipulations of a prejudicial editor, A. J. Henry, whose perspective on the value of her work was further distorted by institutional and national pride, ageism, and sexism. Gender expectations of the time both shaped and severely constrained her subsequent career path. More research is needed on her pioneering contributions, which revolutionize current understandings of women in meteorology.

5. During the first atomic age, physicists believed that ionizing radiation was somehow powering the lower atmosphere and might be responsible for atmospheric phenomena. This story, to my knowledge, has not been told before.

6. Rossby was very much his own man, not merely a Bergen school follower but a formidable theorist and institution builder, mentor of many, and founder of new journals and new approaches. In the short time he spent with Bjerknes, he quickly realized that a mathematical solution to atmospheric motion was out of reach and that the graphical solutions developed in Bergen were the best available. So he took a different course. He stared long and hard at the colored fluids in his rotating tank, watching them perform the integrations that eluded him on paper. He worked on practical projects that made aviation safer, developed upper-air soundings using aircraft and radiosondes, brought Bergen methods to the U.S. Weather Bureau, and trained weather officers during World War II. His passions, however, lay in theory—air

mass and isentropic analysis, conservation of rotational motion (called vorticity), planetary long waves and group velocity, jet streams, and the general circulation of the atmosphere and oceans.

The essence of Rossby's philosophy was to focus on air motion and leave the pressure to adjust itself to the air trajectories. This was Bjerknes's mathematical philosophy, too, and Rossby carried it through to success—at least for the large-scale waves in the upper atmosphere that can be modeled as moving independently of thermodynamic solenoids, friction, radiation, and the water vapor cycle. He wrote simplified but profoundly fundamental equations, not for the messy weather at the surface or for the complex atmosphere as a whole but for the winds aloft and the troughs and ridges in an idealized barotropic atmosphere, treated in the horizontal dimension only, like a thin sheet of paper. Such constraints, however, impose simplifications so great that the models can hardly be called realistic. The hydrodynamics underlying the barotropic model is no longer the complete physical hydrodynamics introduced by Bjerknes. Instead, Rossby emphasized vorticity conservation and its corollary, the Rossby wave equation. Enhancing the model and indispensable for numerical computation was Charney's quasi-geostrophic approximation that filtered out meteorological "noise" such as gravity waves while preserving the large-scale features of the flow. If Joseph-Louis Lagrange went with the flow, Rossby and Charney went with the upper-air flow.

Seeking new challenges, Rossby returned to his native Sweden later in life and turned his primary attention to unsolved atmospheric problems of turbulence, chemical meteorology, and chemical climatology. Rossby made an important turn to global environmental concerns in the later part of his career, fostering an early ethic of environmental responsibility. Rossby never remained in one place for long, and he never wrote a book. There is no biography of Rossby, nor is there a centralized archive of his papers, although several significant biographical articles exist.

7. Three U.S. Weather Bureau chiefs—Charles F. Marvin, Willis Ray Gregg, and Francis W. Reichelderfer—encountered Bergen school methods during their administrations, with the latter two taking the most proactive steps to facilitate their adoption. The first two individuals have been largely ignored in the existing literature.

8. Each of the world wars was devastating to both individuals and to the weather research programs of nations on the losing side. On the other hand, war focused strategic attention and resources on meteorology's

ability to support troop movements, aviation and naval operations, artillery shelling, and poison gas attacks. The hardships Bjerknes experienced during World War I convinced him to move his operations from Germany to Norway; his nation's practical needs refocused his attention on graphical forecasting methods. The isolation Bjerknes and his colleagues experienced during World War II in occupied Norway resulted in devastating delays in the publication of the final volume, *Dynamic Meteorology and Weather Forecasting*. In America, the war emergency gave rise to an urgent and massive training effort led and facilitated by Rossby and Wexler and an expansion of forecasting research worldwide, including the tropics. Ties between civilian and military meteorologists continued after 1945 as did access to equipment for field campaigns. During the Cold War, meteorologists came to see nuclear fallout encircling the world as both a research opportunity and an existential health threat. The satellite era and the space race opened up an era of massive funding and committee work that resulted in the formation of new U.S. and international programs and institutions.

9. Wexler's amply documented career provides a window into the science, technology, and institutional organization of the period 1934 to 1962. He was a leader in air mass analysis, air pollution meteorology, nuclear tracers, and all the new technologies applicable to meteorology. He too articulated a global environmental ethic and a moral role for meteorology and stood opposed to uninformed attempts to modify weather and climate.

10. Claims for weather and climate control, although venerable, appear much more frequently in the literature of meteorology after 1945. The new technologies of digital computing, rocketry, nuclear power, and chemical cloud seeding emboldened some to propose hypothetical, often outrageous, military and commercial applications. Mainstream meteorologists were largely opposed to the excessive promises of rainmakers and climate interventionists. Still, such ideas proliferated and found their way, at least rhetorically, into otherwise serious proposals for research institutions such as NCAR and international collaborations such as the Global Atmospheric Research Program.

11. The interdisciplinary field called "the atmospheric sciences," a loose conglomeration of research specialties, techniques, and institutional affiliations, resulted from the convergence of lives, themes, and

agendas. From about 1957 to 1962, an extremely large number of energetic and insightful individuals contributed to the emergence of atmospheric science. Rossby, Wexler, Lloyd Berkner, Jule Charney, Henry Houghton, Horace Byers, Thomas Malone, Sverre Petterssen, Paul Klopsteg, Alan Waterman, and Walter Orr Roberts stand out in this era. They staffed the committees, lobbied for enhanced funding, and coordinated university, federal, and international activities. The field has expanded since the 1960s becoming part of much larger interdisciplinary umbrellas called, variously, Earth system science, global change science, and integrated environmental assessment.

12. The neo-Laplacian program—to measure the atmosphere with sufficient accuracy and to predict the future with sufficient precision—attracted meteorologists, led by Bjerknes and Rossby, who saw the central problem of the science of meteorology as the prediction of future weather. They worked to improve the practice of weather observation and reporting and incorporated the classical theories of ideal fluid motion of Helmholtz and Kelvin into the equations of hydrodynamics and thermodynamics for real fluids, both in synoptic (Eulerian) form and as Lagrangian trajectories. When the going got too tough, they shifted gears and utilized graphical techniques, simplified mathematics, and approximate computational methods.

Meteorologists were indeed able, using Bergen school methods, to predict fractional rotations of a cyclone over the course of two or three days or the persistence of an anticyclone during fair weather periods. Following Rossby, they were able as well to compute a frog's-eye view of upper-air patterns for about five days into the future. Radar and satellites provided instantaneous bird's-eye views of the weather as it happened. On the other hand, the long-term integration of the hydrodynamic equations, the so-called infinite forecast provided by general circulation models, returned no information at all about specific conditions but generated the statistical features of an unperturbed climate system. The holy grail of intermediate-to long-range forecasts—ten days, thirty days, seasons, or years—eluded them. Two important theorists serve as sobering bookends for the deterministic quest for greater precision and longer and longer forecasts—Henri Poincaré in the early years of the twentieth century and Edward N. Lorenz after 1962. Each, in his own way, placed severe limits on the ability to forecast the future.

Bjerknes's work was fundamental to it all in establishing the scientific foundations of meteorology and setting a neo-Laplacian agenda for the new century. Rossby's theoretical insights, his support for digital computing, and his International Meteorological Institute in Stockholm were direct inspirations for what became atmospheric science. Wexler played key roles in technological development, interagency coordination, and international programs. Throughout the first six decades of the twentieth century, they and their close associates built an interdisciplinary science that inaugurated a new era of atmospheric science, a complex field that has since followed its own set of trajectories, in many ways convergent, in some ways divergent, and perhaps in all ways interesting, strange, and chaotic.

7   Notes

Chapter 1: Introduction

1. Bellamy, *Prospectus of Meteorological Operations*.

2. Fleming, *Meteorology in America*.

3. Archival sources of information on Vilhelm Bjerknes are held by the Carnegie Institution Archives, Washington, DC; the Nasjonalbiblioteket, Oslo; and the Statsarkivet and the Geophysical Institute Archives, Bergen. Sources on Carl-Gustaf Rossby were compiled from a large number of sources. The Harry Wexler Papers comprise a massive forty-four-box collection in the Manuscript Division of the U.S. Library of Congress. His private papers are held by his family.

4. Multhauf, "Review of *A History of the United States Weather Bureau* by Donald R. Whitnah," 333.

5. Christie, "Aurora, Nemesis and Clio."

6. U.S. National Academy of Sciences, Committee on Atmospheric Sciences, "Meteorology on the Move."

7. Evangelista Torricelli to Michelangelo Ricci, June 11, 1644, as quoted in Walker, *Ocean of Air*, 18.

8. Abbe, *Mechanics the Earth's Atmosphere*, multiple volumes; Kutzbach, *Thermal Theory*.

9. Newton, "Roundtable Discussion about C. G. Rossby."

10. Roulstone and Norbury, *Invisible in the Storm*, focuses on the nontechnical, mathematical history of meteorology; Edwards, *A Vast Machine*, is a broad survey of weather- and climate-related information infrastructures and technology; R. M. Friedman, *Appropriating the Weather*, examines in detail the life and work of Vilhelm Bjerknes to the 1920s; Meyer, *Americans and Their Weather*, is a social history with two twentieth-century chapters; Nebeker, *Calculating the Weather*, and Harper,

Weather by the Numbers, emphasizes computation to the exclusion of other factors; Whitnah, *History of the United States Weather Bureau*, focuses on bureaucracy rather than research; Bates and Fuller, *America's Weather Warriors*, examines the role played by the U.S. military. Historical accounts by practitioners Bergeron, Platzman, and Smagorinsky are some of the most useful of this genre.

Chapter 2: Bjerknes

1. Doudan, *Mélanges et lettres*, 3:164.

2. This section is based on V. Bjerknes, *C. A. Bjerknes*. The copy borrowed from Oregon State University included the following handwritten inscription to Erwin Schrödinger: "A book about prehistoric physics. E. Schrödinger dedicated by V. Bjerknes in memory of our days spent together in Rome 1937" (at the Papal Academy of Sciences).

3. "Telephone at the Paris Opera."

4. A. P. Trotter, as quoted in Hughes, *Networks of Power*, 50.

5. Pihl, "Bjerknes, Carl Anton"; C. A. Bjerknes, "Hydrodynamiske Analogier."

6. Euler, *Letters . . . to a German Princess*, letter 19, June 14, 1760.

7. V. Bjerknes, "Hvordan Bergensskolen ble til."

8. Forbes, "Hydrodynamic Analogies."

9. V. Bjerknes, "Nyere hydrodynamiske Undersøgelser."

10. R. M. Friedman, *Appropriating the Weather*; R. M. Friedman, "Bjerknes, Vilhelm."

11. V. Bjerknes, *Vorlesungen über hydrodynamische Fernkräfte*; V. Bjerknes, *Fields of Force*; V. Bjerknes, "Hvordan Bergensskolen ble til"; Darrigol, *Worlds of Flow*, 177, 187–197.

12. Poincaré, *Les méthodes nouvelles*; Devik, cited in *In Memory of Vilhelm Bjerknes*, 9; "Publications by Vilhelm Bjerknes," 26–37.

13. Hertz, *Principles of Mechanics*.

14. V. Bjerknes, "Om elektricitetsbevregelsen."

15. Honoria Bjerknes Hamre, personal communication to the author, June 12, 2014.

16. Gold, "Vilhelm Friman Koren Bjerknes," 305.

17. Bergeron, Devik, and Godske, *In Memory of Vilhelm Bjerknes*.

18. Helmholtz, "On the Integrals of the Hydro-dynamic Equations"; Helmholtz, "On Atmospheric Movements."

19. Wilczek, "Beautiful Losers."

20. V. Bjerknes, "Course of Twenty-nine Lectures," lecture 14; V. Bjerknes, *Vorlesungen über hydrodynamische Fernkräfte;* On Helmholtz and hydrodynamics, see Darrigol, *Worlds of Flow*, 145–182.

21. Fysiska Sällskapets i Stockholm, Protokollsbok 1 (1891–1898) and 2 (1899–1915). Kungliga Vetenskapsakademien Archives, Stockholm.

22. V. Bjerknes, "Über einen hydrodynamischen Fundamentalsatz."

23. Gold, "Vilhelm Friman Koren Bjerknes," 306.

24. V. Bjerknes, "Circulatory Movements in the Atmosphere."

25. Abbe to V. Bjerknes, December 9, 1898, NBO, Brevs 469B; V. Bjerknes, "Dynamische Princip"; V. Bjerknes, "Dynamic Principles"; V. Bjerknes, "Über die Bildung"; V. Bjerknes, "Circulatory Movements in the Atmosphere"; Ekholm, *Étude des conditions météorologiques.*

26. Ellingsen, "Varme havstrømmer og kald krig."

27. Helland-Hansen and Nansen, *Norwegian Sea*; V. Bjerknes et al., *Hydrodynamique physique*, 842–843; R. M. Friedman, *Appropriating the Weather*, 42–44; Mills, *Fluid Envelope*, 106–110.

28. Miller, "Science and Private Agencies," 191–221, cited on 215.

29. Carnegie, "Trust Deed," 12.

30. Ibid., 11.

31. Cleveland Abbe to Charles D. Walcott, July 14, 1902, CIS Archives; *Carnegie Institution Yearbook* 1 (1902): 76–82, quote from 79.

32. Hann to Walcott, July 17, 1903; Bigelow to Walcott, May 12, 1902; Rotch and Moore, cited in Abbe to Walcott, July 14, 1902, CIS Archives; Brown, *Centennial History;* Trefil and Hazen, *Good Seeing.*

33. Abbe, "Physical Basis of Long-Range Weather Forecasts," 561; Willis and Hooke, "Cleveland Abbe."

34. Abbe, "Needs of Meteorology," 181–182.

35. V. Bjerknes, "Cirkulation relativ zu der Erde," 108.

36. V. Bjerknes, "Das Problem der Wettervorhersage." Bjerknes had developed these ideas in 1903 in a lecture presented to the Stockholm Physics Society and in subsequent discussions. R. M. Friedman, *Appropriating the Weather*, 53; Doyle et al., "Mesoscale Modeling," 8.

37. Laplace, "Philosophical Essay on Probabilities"; cf. Roulstone and Norbury, *Invisible in the Storm.*

38. V. Bjerknes, *Fields of Force*; V. Bjerknes et al., *Physikalische Hydrodynamik*, "Bibliographie mit historischen Erläuterungen."

39. V. Bjerknes, *Fields of Force*, 129–130, 153–154.

40. V. Bjerknes, *Fields of Force*, 153–154; Sartz, "Norway's Contributions," 540.

41. V. Bjerknes to Honoria Bjerknes, November 27, 30, December 12, 25, 1905, and January 6, 1906, NBO, Familien Bjerknes, Brevs 469C.

42. Bjerknes to Honoria, December 17, 19, 1905, Familien Bjerknes, NBO, Familien Bjerknes, Brevs 469C.

43. V. Bjerknes, "Hvordan Bergensskolen ble til," 10; cf. Bonacina, "The Great Problem of Meteorology."

44. Lavine, *First Atomic Age*, 23.

45. Dadourian, "On the Constituents."

46. Bumstead, "Atmospheric Radio-activity"; Eve, "On the Ionization of the Atmosphere," 27.

47. Curie, "Radioactive Substances."

48. Arrhenius, *Worlds in the Making*, 69.

49. Carlson, "Discovery of Cosmic Rays."

50. Hess, "Unsolved Problems in Physics."

51. V. Bjerknes, "Abstract of Application, Meteorology," February 7, 1906, CIS Archives.

52. Bjerknes to the Trustees of the Carnegie Institution, February 7, 1906, reply March 26, 1906, CIS Archives.

53. Bjerknes to Woodward, April 12 and May 7, 1906, CIS Archives; *Carnegie Institution Yearbook* 5 (1907): 212.

54. V. Bjerknes, "Synoptical Representation of Atmospheric Motions."

55. V. Bjerknes, "Meteorologie als Exacte Wissenshaft."

56. Bjerknes to Woodward, June 25, 1912, CIS Archives.

57. V. Bjerknes, "Meteorologie als Exacte Wissenshaft"; V. Bjerknes, "Meteorology as an Exact Science."

58. Ibid.

59. Bjerknes to Woodward, May 6, 1913, CIS Archives. A bar is the metric unit of pressure, approximately equal to one atmosphere; a millibar is one-one-thousandth of a bar.

60. Woodward to Bjerknes, May 19, 1913, CIS Archives; *Carnegie Institution Yearbook* 12 (1913): 298–299.

61. Woodward to Bjerknes, January 19, 1914, CIS Archives.

62. "New Daily Weather Map." The U.S. Army Signal Office issued the first international weather maps of the Northern Hemisphere between 1872 and 1889.

63. *Nature* 92 (February 20, 1914): 715–716.

64. Abbe, "Weather Map on the Polar Projection."

65. Marvin, "Northern Hemisphere Map Interrupted"; U.S. Weather Bureau, *Report of the Chief*, 1913–1914, 63rd Cong., 3rd Sess., H. Doc. 1554, p. 10.

66. V. Bjerknes to Woodward, July 29, 1914, CIS Archives.

67. V. Bjerknes to Woodward, August 17, 1914, CIS Archives.

68. Woodward to V. Bjerknes, September 4 and 8, 1914, CIS Archives.

69. Bergeron, Devik, and Godske, *In Memory of Vilhelm Bjerknes*.

70. V. Bjerknes to Woodward, July 24 and December 22, 1915, CIS Archives.

71. V. Bjerknes to Woodward, September 26, 1914, CIS Archives.

72. J. Bjerknes, "Über die Fortbewegung der Konvergenz"; Eliassen, "Jacob Aall Bonnevie Bjerknes."

73. V. Bjerknes to Woodward, August 16, 1917, CIS Archives.

74. R. M. Friedman, *Appropriating the Weather*, 101 n. 22, 102.

75. V. Bjerknes, "Geofysiker motet i Göteborg."

76. J. Bjerknes, "On the Structure of Moving Cyclones."

77. Bergeron, "Methods in Scientific Weather Analysis and Forecasting."

78. J. Bjerknes and Solberg, "Meteorological Conditions"; J. Bjerknes and Solberg, "Life Cycle of Cyclones"; Shapiro and Grønås, *Life Cycle*.

79. V. Bjerknes to Woodward, July 7, 1920, CIS Archives, emphasis in the original.

80. *Carnegie Institution Yearbook* 18 (1919): 351.

81. V. Bjerknes to Woodward, August 9, 1918, CIS Archives.

82. Woodward to V. Bjerknes, September 12, 1918, CIS Archives.

83. Woodward to Marvin, March 26, 1919 and reply, March 27, 1919, CIS Archives.

84. V. Bjerknes, "Possible Improvements in Weather Forecasting"; Marvin, "Status and Problems."

85. V. Bjerknes, "Meteorology of the Temperate Zone"; R. M. Friedman, *Appropriating the Weather*, 189–195.

86. V. Bjerknes to Woodward, January 26, 1921, CIS Archives.

87. V. Bjerknes, "Cirkulation relativ zu der Erde." The Ferrel circulation is named for the American geophysicist William Ferrel.

88. "A Conference in Bergen"; Davies, "Emergence of the Mainstream Cyclogenesis Theories"; R. M. Friedman, *Appropriating the Weather*, 196.

89. "Meeting of the International Commission."

90. Exner, *Dynamische Meteorologie,* foreword; Ficker, "Polarfront, Aufbau"; Schwerdtfeger, "Comments on Tor Bergeron's Contributions," 501–502.

91. Exner, "Anschauungen über kalte und warme Luftströmungen."

92. Coen, *Vienna in the Age of Uncertainty,* 289.

93. Coen, *Vienna in the Age of Uncertainty,* 282-292; Coen, "Scientific Dynasty," 293; Coen, "Scaling Down," 117.

94. R. M. Friedman, *Appropriating the Weather,* 199.

95. Khromov, *Einführung in die synoptische Wetteranalyse.* The 532-page tome contains numerous maps, a very thorough bibliography, a name index, and a topical index.

96. R. M. Friedman, *Appropriating the Weather,* 199; Anders Persson, personal communication, December 7, 2013.

97. Hong, *Wireless,* 191.

98. Aitken, *Continuous Wave,* 250–251.

99. Douglas, *Inventing American Broadcasting,* xv.

100. U.S. Weather Bureau, *Report of the Chief,* 1920–1921, H. Doc. 226, 67th Cong., 2nd Sess., p. 11; U.S. Weather Bureau, *Report of the Chief,* 1921–1922, H. Doc. 541, 67th Cong., 4th Sess., p. 12.

101. U.S. Weather Bureau. *Report of the Chief,* 1921–1922, H. Doc. 541, 67th Cong., 4th Sess., pp. 13–14.

102. Gold, "Vilhelm Friman Koren Bjerknes."

103. *Oxford English Dictionary,* s.v. "ionosphere."

104. Appleton, "Ionosphere."

105. Mitra, "General Aspects of Upper Atmospheric Physics."

106. "Fellowship in Meteorology."

107. VpV Archives, Udgaaende Korrespondanse, August 28, 1920.

108. "Bergen Geophysical Institute."

109. Anne Louise Beck to Prof. and Mrs. Bjerknes, September 9, 1921, NBO, Brevs 469B.

110. Beck, "An Application of the Principles."

111. Beck, "Earth's Atmosphere as a Circular Vortex."

112. Godske, cited in *In Memory of Vilhelm Bjerknes*, 23.

113. Lou McNally to the author, personal communication, July 10, 2014.

114. Bailey, *Santa Rosa Junior College*, 19.

115. Ibid., 24–33.

116. *Patrin*, 1930.

117. *Oak Leaf* (September 26, 1941).

118. Bjerknes to Merriam, July 25, 1922; Walter M. Gilbert to Marvin, September 1, 1922; Marvin to Gilbert, September 14, 1922 and reply, September 28, 1922, CIS Archives.

119. Toronto *Globe*, August 12, 1924.

120. V. Bjerknes, "Forces Which Lift Aeroplanes"; "British Association Meeting at Toronto."

121. Toronto *Globe*, August 12, 1924; V. Bjerknes, "Forces Which Lift Aeroplanes."

122. J. Bjerknes, "The Importance of Atmospheric Discontinuities for Practical and Theoretical Weather Forecasting," MS lecture, Toronto, 1924, Statsarkivet i Bergen.

123. "British Association Meeting at Toronto"; V. Bjerknes to Honoria, August 13–September 24, 1924, NBO, Familien Bjerknes, Brevs 469C; V. Bjerknes to Merriam, May 28, 1924, CIS Archives.

124. V. Bjerknes to Merriam, August 17, 1924, CIS Archives.

125. V. Bjerknes, "Course of Twenty-nine Lectures," Lecture 13; Helmholtz, "Wirbelstürme und Gewitter."

126. V. Bjerknes, "Course of Twenty-nine Lectures," Lecture 13.

127. Ibid.; V. Bjerknes, "Solar Hydrodynamics."

128. V. Bjerknes to Honoria, December 12, 1924, NBO, Familien Bjerknes, Brevs 469C.

129. "Lectures before the Carnegie Institution by Prof. V. Bjerknes, Bergen, Norway, Dec. 27, 1924" (to be given early January), CIS Archives.

130. V. Bjerknes to Gilbert, October 30 and reply November 5, 1924; V. Bjerknes to Gilbert, November 26 and reply, December 3, 1924; V. Bjerknes to Gilbert, January 17, 1925, CIS Archives; V. Bjerknes to Honoria, Jan. 23, 1925, NBO, Familien Bjerknes, Brevs 469C.

131. V. Bjerknes et al., *Physikalische Hydrodynamik*.

132. Richardson, *Weather Prediction by Numerical Process*; Ashford, *Prophet—or Professor?*; Lynch, *Emergence of Numerical Weather Prediction*.

133. V. Bjerknes to Merriam, January 4, 1925, CIS Archives.

134. V. Bjerknes to Merriam, May 28, 1924, CIS Archives; R. M. Friedman, *Appropriating the Weather*, 233–235; V. Bjerknes, "Hvordan Bergensskolen ble til."

135. V. Bjerknes to Merriam, August 15, 1928, CIS Archives.

136. Ibid.

137. V. Bjerknes et al., *Hydrodynamique physique*.

138. V. Bjerknes et al., *Physikalische Hydrodynamik*, "Bibliographie mit historischen Erläuterungen."

139. V. Bjerknes to Merriam, October 16, 1933, CIS Archives.

140. V. Bjerknes to Merriam, February 22, 1935, CIS Archives.

141. V. Bjerknes to Merriam, June 21, 1935, and January 31, 1936, CIS Archives; J. Bjerknes, "Practical Examples of Polar-Front Analysis over the British Isles"; Gold, "Fronts and Occlusions"; Simpson, "Weather Forecasting"; Shaw, *Drama of Weather*, 245–253.

142. V. Bjerknes to Merriam, July 3, 1937, CIS Archives.

143. V. Bjerknes, "Leipzig-Bergen."

144. V. Bjerknes to Vannevar Bush, March 12, 1939; Bush to Gilbert and reply, CIS Archives.

145. Bush to Rossby, May 6, 1940, CIS Archives.

146. Rossby to Bush, May 15, 1940, CIS Archives.

147. Ibid.

148. Petterssen, *Weathering the Storm*.

149. J. Bjerknes to Bush, June 17, 1940, CIS Archives.

150. J. Bjerknes to Gilbert, September 27, 1940, CIS Archives.

151. J. Bjerknes to Bush, August 21, 1945; Bush to V. Bjerknes, October 22, 1945, CIS Archives; V. Bjerknes, "Preface," in Godske et al., *Dynamic Meteorology*.

152. Sawyer, "Review of *Dynamic Meteorology*."

153. V. Bjerknes, "Hvordan Bergensskolen ble til"; V. Bjerknes, "Dynamics and Electromagnetism"; *Carnegie Institution Yearbook* 51 (1951): 18.

154. V. Bjerknes, "Cirkulation relativ zu der Erde," 108; V. Bjerknes, "Das Problem der Wettervorhersage," 1–2.

155. Bergeron, cited in *In Memory of Vilhelm Bjerknes*, 18.

156. Gold, "Vilhelm Friman Koren Bjerknes," 313–314.

157. "Man's Milieu," 73; Bergeron and Devik, cited in *In Memory of Vilhelm Bjerknes*, 18, 25.

Chapter 3: Rossby

1. Rossby, "Comments on Meteorological Research," 32.

2. "Man's Milieu," 68.

3. Lewis, "Carl-Gustaf Rossby"; Kutzbach, "Rossby."

4. Rossby, "Biographical Sketch"; Rossby's bibliography is in Byers, "Carl-Gustaf Arvid Rossby," 265–270. His preserved correspondence is rather thin with no central repository.

5. *Vervarslinga på Vestlandet 25 år Festskrift*; Bergeron, "Young Carl-Gustaf Rossby," 53; Phillips, "Carl-Gustaf Rossby," 1097.

6. Rossby, "Den nordiska aerologiens arbetsuppgifter."

7. Van Bebber, *Lehrbuch der Meteorologie*, 309.

8. Kvinge, "*Conrad Holmboe* Expedition."

9. Berry, Bollay, and Beers, *Handbook of Meteorology*, 561.

10. Rossby, "Meteorologiska Resultat"; Phillips, "Carl-Gustaf Rossby," 1099–1100.

11. Bergeron, "Young Carl-Gustaf Rossby," 51; Byers, "Carl-Gustaf Arvid Rossby," 253; Phillips, "Carl-Gustaf Rossby," 1101.

12. On Anne Louise Beck, see chapter 2; J. Bjerknes and Giblett, "Analysis of a Retrograde Depression."

13. Helmholtz, "On the Integrals of the Hydro-dynamic Equations," 57; Koenigsberger, *Hermann von Helmholtz*.

14. Rossby, "On the Origin of Travelling Discontinuities"; Rossby, "On the Solution of Problems." In the late 1940s, Dave Fultz at Chicago produced experimental results linked to numerical analysis.

15. Rossby and Weightman, "Application of the Polar-Front Theory"; Weightman and Rossby, "Polar Fronts in the United States"; Rossby, "Theory of Atmospheric Turbulence."

16. Guggenheim, "[First] Report . . . 1926–1927," ii; Guggenheim, "Final Report . . . 1929," i; Hallion, *Guggenheim Contribution*.

17. Smith, *Airways*, 94–102.

18. Bates, "Formative Rossby-Reichelderfer Period," 595.

19. Reichelderfer, *Norwegian Methods*.

20. Byers, "Carl-Gustaf Arvid Rossby"; "Lindbergh's Mexican Trip Deferred by Bad Weather," 1; Reichelderfer, "The Atmosphere."

21. Bowie, *Weather and the Airplane*, 5–6.

22. U.S. Weather Bureau, *Report of the Chief*, 1932–33, H. Doc. 121, 73rd Cong., 2nd Sess.

23. Marvin to S. W. Stratton, April 4, 1927, and reply April 8, 1927, MIT Office of the President, AC 13, B29, F493; Humphreys to G. K. Burgess, March 8, 1927, idem.

24. Stratton to Marvin, April 16, May 3, May 5, 1927, and reply April 21; Stratton to Norton, May 5, 1927, MIT Office of the President, AC 13, B29, F493.

25. Rossby to Stratton, July 2, 1927, and reply July 11, 1927; Stratton to Norton, July 7, 1927, MIT Office of the President, AC 13, B29, F493.

26. A. T. Church to Stratton, March 20, 1928, and reply May 11, 1928, MIT Office of the President, AC 13, B29, F493. The six navy students were Clifford M. Alvord, Vernon O. Clapp, Walter E. Gist, Howard B. Hutchinson, Robert H. Smith, and Herbert M. Wescoat.

27. Stratton to Guggenheim, May 26, 1928, and reply; Warner to Stratton, June 14, 1928, MIT Office of the President, AC 13, B29, F493.

28. Guggenheim to Stratton, June 18, 1938, and reply June 21, MIT Office of the President, AC 13, B29, F493; Massachusetts Institute of Technology, "Brief History of the Department of Meteorology."

29. Rossby, "Outline of a Research Program," *MIT President's Report*, 1928.

30. Rossby to Stratton, November 5, 1928; Rossby, "Outline of a Research Program for the Meteorology Department, Massachusetts Institute of Technology," MIT Office of the President, AC 13, B29, F493.

31. Willett, "Dynamic Meteorology."

32. Rossby to Reichelderfer, November 22, 1928, U.S. Library of Congress, Reichelderfer Papers, box 3; Rossby to Stratton, February 26, 1929, MIT Office of the President, AC 13, B29, F493; Rossby, "First Annual Report."

33. Rossby to Stratton, March 15 and May 17, 1929, MIT Office of the President, AC 13, B29, F493. For more on fog research at MIT, see Fleming, *Fixing the Sky*, 122–127.

34. Namely, the Subcommittee on Meteorological Problems of the Committee on Air Navigation of the National Advisory Committee for Aeronautics and the Committee on Meteorology of the National Research Council.

35. Stratton, undated note, early 1929, MIT Office of the President, AC 13, B29, F494.

36. Stratton to Rossby June 30, 1930, MIT Office of the President, AC 13, B29, F494.

37. Stratton to Rossby, July 8, 1929, MIT Office of the President, AC 13, B29, F494.

38. Rossby to Stratton, March 10, 1930, MIT Office of the President, AC 13, B29, F494.

39. Phillips, "Carl-Gustaf Rossby," 1103.

40. *MIT President's Report 1932.*

41. Rossby to Compton, December 11, 1931, enclosing memorandum of agreement between MIT and Woods Hole, MIT Office of the President, AC 4, B147, F11.

42. Berry, Bollay, and Beers, *Handbook of Meteorology*, 563.

43. London, "Bernhard Haurwitz"; Haurwitz, "Meteorology in the Twentieth Century."

44. John A. Fleming to Walter W. Gilbert, November 28, 1933, CIW Archives.

45. Bergeron, "Young Carl-Gustaf Rossby," 55; Phillips, "Carl-Gustaf Rossby," 1103–1104; Byers, "Carl-Gustaf Arvid Rossby."

46. Bergeron, "Young Carl-Gustaf Rossby," 54.

47. Henry, "Weather Forecasting," 14–15.

48. *MIT President's Report 1933.*

49. Millikan et al., "Preliminary Report of the Special Committee."

50. Hinsdale, "Charles Frederick Marvin."

51. Brooks, "Free Air Conditions"; Gregg, "Meteorological Service for Airways"; Gregg, "Standard Atmosphere"; American Meteorological Society, *Glossary of Meteorology*; Humphreys, "Willis Ray Gregg."

52. Gregg, "Daily Record of Appointments, Visits, Etc. 1934–1938," typescript, Reichelderfer Records, NARA, box 1.

53. Gregg, "Weather Bureau and the Nation's Business"; Gregg, "Progress in Development of the U.S. Weather Service"; U.S. Weather Bureau, *Report of the Chief 1933–34*, H. Doc. 11, 74th Cong., 1st Sess.; Rossby, "Annual Report," *MIT President's Report 1934*.

54. Braham and Malone, "Horace Robert Byers."

55. Bates, "Formative Rossby-Reichelderfer Period," 599–600.

56. Petterssen, *Weathering the Storm*; Pettersson's lecture notes, Wexler Papers 37.

57. Weightman, "Forecasting from Synoptic Weather Charts."

58. Rossby, "Memorandum Concerning a Course in Weather and Communications Service Especially Adapted to the Needs of Air Transport Companies, March 16, 1929," MIT Office of the President, AC 13, B29, F494.

59. Gregg, "Radio-Meteorographs for Unmanned Balloons," typescript, December 29, 1936, Material from Previous Chiefs, Reichelderfer Records, NARA, box 1; Gregg, "Progress in Weather Forecasting," 407, 412; Lange, "Radiometeorographs"; Wenstrom, "Radiometeorography."

60. Jeon, "Flying Weather Men." Gregg appears in this work as "Marvin's successor" or "Reichelderfer's predecessor," but his own accomplishments are not discussed.

61. DuBois, Multhauf, and Ziegler, *Invention and Development of the Radiosonde*, abstract and 25–27.

62. Ibid., cover image caption, 34–35; Curry and Pyle, *Encyclopedia of Atmospheric Sciences*, 1912–1913.

63. Gregg, "A Visit to the U.S.S.R.: An Unofficial Account of an Official Trip," Material from Previous Chiefs, Reichelderfer Records, NARA, box 1.

64. Devereaux, "Meteorological Service of the Future"; Mindling, "Raymete and the Future"; Fleming, *Fixing the Sky*, 133–134.

65. Rossby to Compton, June 9, 1940, MIT Office of the President AC 4, B187, F2.

66. Rossby to Compton, Oct. 15, 1939, MIT Office of the President AC 4, B187, F2.

67. Ibid.

68. Compton to Rossby, November 3, 1939, MIT Office of the President, AC 4, B187, F2.

69. Rossby, "Scientific Basis of Modern Meteorology."

70. Leighly to Rossby, November 9. 1939, MIT Office of the President, AC 4, B187, F2.

71. Rossby, "Scientific Basis of Modern Meteorology," 652.

72. Platzman, *Conversations with Jule Charney*, 58–59, 151–152; see also Lindzen, Lorenz, and Platzman, *The Atmosphere*.

73. The author first learned this in a course given by Haurwitz, "Geophysical Hydrodynamics."

74. Rossby and collaborators, "Relation between Variations"; Rossby, "Planetary Flow Patterns."

75. Platzman, *Conversations with Jule Charney*, 64; Rossby, Oliver, and Boyden, "Weather Estimates"; Rossby, "Specific Evaluation."

76. Rossby to Compton, June 9, 1940, MIT Office of the President, AC 4, B187, F2.

77. Byers, "Founding of the Institute of Meteorology."

78. Lewis, "Cal Tech's Program." The Caltech program was established by Theodore von Kármán in 1933 with the support of the Guggenheim fund and ran for fifteen years. New York University, "Department of Meteorology."

79. Curriculum of the "AAFTTC Weather Training Program, Class 'A'," MIT Office of the President, AC 4, B148, F1; Byers, "Recollections of the War Years." Participating meteorology schools, in alphabetical order, included Amherst College, Bowdoin College, Brown University, California Institute of Technology, Carleton College, Denison University, Grand Rapids Junior College, Hamilton College, Haverford College, Kenyon College, Massachusetts Institute of Technology, New York University, Pomona College, Reed College, State University of Iowa, University of California at Berkeley, University of Michigan, University of Minnesota, University of New Mexico, University of North Carolina at Chapel Hill, University of Oregon, University of Southern California, University of Virginia, University of Washington, University of Wisconsin, Vanderbilt University, and Washington University at St. Louis. See MC511 University Meteorology Committee, MIT Institute Archives.

80. Riehl, *Tropical Meteorology*; Berry, Bollay, and Beers, *Handbook of Meteorology*, 763–803.

81. Lewis, Fearon, and Klioforth, "Herbert Riehl"; Petterssen, *Weathering the Storm*; Fleming, "Sverre Petterssen."

82. Petterssen, *Weathering the Storm*, 266–273.

83. *Flight and Aircraft Engineer* 38 (July 4, 1940), 6.

84. *Flight and Aircraft Engineer* 61 (April 25, 1952).

85. Rossby to Don N. Yates, May 19, 1945, MIT Office of the President, AC 4, B187, F2.

86. Ibid.

87. Ibid.

88. Ibid.

89. Byers, "Recollections of the War Years"; Spiegler, "History of Private Sector Meteorology"; Cartwright and Sprinkle, "History of Aeronautical Meteorology."

90. Rossby, "Horizontal Motion in the Atmosphere"; Rossby, "On the Propagation of Frequencies"; Sverdrup, Johnson, and Fleming, *The Oceans;* Sverdrup, *Oceanography for Meteorologists.*

91. Rossby and Staff, "On the General Circulation."

92. Platzman, *Conversations with Jule Charney*, 34–35.

93. Ibid., 11–18.

94. Ibid., 19–24, 30; Charney, "Radiation."

95. Platzman, *Conversations with Jule Charney*, 38–39.

96. Ibid., 47–48.

97. Ibid., 55.

98. Ibid., 56.

99. Ibid., 27, 36–37, 45–47; Charney, "Dynamics of Long Waves."

100. Platzman, *Conversations with Jule Charney,* 56; Charney, "On a Physical Basis for Numerical Prediction."

101. Ibid., 59.

102. Platzman, *Conversations with Jule Charney*, 61–63.

103. Charney to Rossby, March 19, 1947, Charney Papers B14 F460; Aspray, *John Von Neumann.*

104. Platzman, *Conversations with Jule Charney*, 78–81.

105. Ibid., 68–70; Lynch, *Emergence of Numerical Weather Prediction.*

106. Rossby to Charney, January 9, 1949, Charney Papers, B14 F460.

107. Charney to Philip Thompson, November 4, 1947, copy in Platzman, *Conversations with Jule Charney*, 165–166.

108. Persson, "Early Operational Numerical Weather Prediction outside the USA . . . Part I."

109. J. Bjerknes, "Half a Century of Change." Rossby's Stockholm period is well documented in Meteorologiska Institutionen, Professor Carl-Gustaf Rossbys Korrespondens, 1948–1957, Stockholm University.

110. Willett, "C.-G. Rossby."

111. Rossby's extracurricular activities included lectures by Tor Heyerdahl on the Kon-Tiki expedition and Hans Ahlmann honoring Finn Ronne's polar exploits and performances of *The Marriage of Figaro* and *Death of a Salesman*. He was reading Gaston Diehl, *Vermeer* (1948); James F. Byrnes, *Speaking Frankly* (1947); Geoffrey Gorer, *The American People: A Study in National Character* (1948); and P.M.S. Blackett, *Military and Political Consequences of Atomic Energy* (1948).

112. Persson, "Early Operational Numerical Weather Prediction outside the USA . . . Part I."

113. Rossby, C.-G., "Note on Activity"; Persson, "Early Operational Numerical Weather Prediction outside the USA . . . Part I."

114. Bolin, "Carl-Gustaf Rossby: The Stockholm Period."

115. Yaskell, "From Research Institution to Astronomical Museum."

116. Rossby to Charney, January 7, 1951, Charney Papers, B14, F459.

117. Rossby to Platzman, May 8, 1949, Charney Papers, B14, F460; Rossby to Charney, September 25, 1948, Charney Papers, B14, F460; Charney to Rossby December 17, 1951, Charney Papers, B14, F459.

118. Rossby, "Note on Cooperative Research Projects."

119. Ibid., 215.

120. "Man's Milieu," 74; Eriksson, "Report on an Informal Conference"; Eriksson, "Report on the Second Informal Conference"; Neumann, "The Fourth Annual Conference."

121. Rothschild, "Burning Rain"; Rossby and Egner, "On the Chemical Climate."

122. Fleming, *Callendar Effect*; Fonselius and Koroleff, "Microdetermination of CO_2"; Bohn, "Concentrating on CO_2."

123. Bert Bolin to John Lewis, October 7, 1991, Lewis Personal Research Collection.

124. Bolin and Eriksson, "Changes in the Carbon Dioxide Content"; Fleming, *Climate Change and Anthropogenic Greenhouse Warming*.

125. Rossby, "Current Problems in Meteorology."

126. Ibid.; Fleming, *Fixing the Sky*.

127. Bergeron, "Young Carl-Gustaf Rossby," 51; Rossby to Charney, April 16, 1952, Charney Papers, B14, F459.

128. Rossby to Charney, July 13, 1956, Charney Papers, B14, F459.

129. Charney, "Death of Rossby," August 20, 1957, Charney Papers, B14, F460.

130. Lewis, "Carl-Gustaf Rossby," 1433.

131. Bolin, "Carl-Gustaf Rossby in Memoriam."

132. Bert Bolin to John Lewis, October 7, 1991, Lewis Personal Research Collection.

133. Rossby, "Current Problems in Meteorology."

134. A list of students and colleagues who were mentored by Rossby includes Hurd Willet, Chaim Pekeris, Horace Byers, Jerome Namias, Raymond Montgomery, Harry Selwell, Richie Simmers, Harry Wexler, Philip Clapp, Henry Houghton, Morris Neiburger, Victor Starr, Jule Charney, Arnt Eliassen, David Fultz, Reid Bryson, Hsio-Lan Kuo, George Platzman, Tu-Cheng Yeh, Koo Chen-Chao, Norman Phillips, George Cressman, Yi-Ping Hsieh, Joanne Malkus (Simpson), Chester Newton, John Freeman, Daniel Rex, Bert Bolin, Edward N. Lorenz, Erik Eriksson, Aksel Wiin-Nielsen, and Bo Döös. Numerous international scholars spent significant time in Stockholm, including, notably, from war-torn Europe, Hans Ertel, Karl Hinkelmann, and Ernst Kleinschmidt from Germany; Fritz Defant from Austria; Frank Ludlam from Great Britain; Erik Palmén from Finland; and Jacques Van Mieghem from Belgium. Lewis, "Carl-Gustaf Rossby"; Bolin to Lewis, October 7, 1991, Lewis Personal Research Collection.

135. "Man's Milieu," 79.

Chapter 4: Wexler

1. Harvard College Class of 1932, *Twenty-fifth Anniversary Report*, 1211, Wexler Papers 19. Parts of this chapter are based on material published in Fleming, "Polar and Global Meteorology"; Fleming, "Beyond Prediction"; and Fleming, *Fixing the Sky*.

2. Jerrold Wexler to Susan Schneider, July 30, 2010; Jerrold Wexler to all Wexler descendants, June 1, 1991, Wexler Family Papers.

3. Biographical information from the Wexler Papers, Library of Congress; Yalda, "Wexler, Harry"; Durfee High School, *Record,* 1928; Harvard College Class of 1932, *Twenty-fifth Anniversary Report*, 1211, Wexler Papers 19.

4. "Report on Meteorological Conference at M.I.T., Feb. 26–March 2, 1934," Wexler Papers 19.

5. Wexler, "Comparison of the Linke and Ångstrom Measures"; Wexler and Haurwitz, "Trübungsfaktoren"; Wexler, "Turbidities of American Air Masses."

6. Harry to Hannah, November 24, 1934, Wexler Family Papers.

7. Wexler, "Deflections Produced in a Tropical Current."

8. Wexler, "Analysis of a Warm-Front"; Wexler, "Cooling in the Lower Atmosphere."

9. Wexler, "Formation of Polar Anticyclones"; Wexler, "Observations of Nocturnal Radiation."

10. Wexler, "Memorandum for Dr. Rossby Concerning Lowell Observatory and Bankhead-Jones Funds," August 25, 1939, Wexler Papers 1; Elsaesser, *Heat Transfer by Infrared Radiation.*

11. Wexler, "Lectures I and II at Caltech," March 3 and 8, 1939, Wexler Papers 15.

12. Wexler, "Observed Transverse Circulations in the Atmosphere."

13. Wexler, U.S. Civil Service Commission Application Form, September 1940, Wexler Papers 37.

14. Wexler, "The Weather"; Wexler et al., "Weather Elements."

15. Wexler to Rossby, January 28, 1943, Wexler Papers 2.

16. Wexler, "Structure of the September 1944 Hurricane."

17. Wexler, "Note to Miles Harris," January 28, 1962, Wexler Papers 14; Wood, "Flight into the September, 1944 Hurricane"; Wexler, "Structure of the September, 1944 Hurricane."

18. Wexler to Adjutant General, August 17, 1945, Wexler Papers 2.

19. Wexler, "Description of Work," Wexler Papers 37; Wexler, "Meet the Division Chiefs."

20. Wexler "Antarctic Diary," May 23, June 3, 1957, Wexler Papers 27; "New Horizons in Science: Weather Satellites," September 22, 1961, U.S. Information Agency, Filename: Wexler 306_10-140 VOA 1961 New Horizons.wav, NARA.

21. Wexler, "Meteorological Aspects of Atomic Radiation"; Wexler, preface to *Meteorology and Atomic Energy.*

22. Wexler and Sheridan, "Study of Barograph Records."

23. Uda, "Meteorological Conditions Related to the Atomic Bomb Explosion at the Hiroshima City."

24. Fleming, *Fixing the Sky*; Machta, "Oral History Interview."

25. Machta and Harris, "Effects of Atomic Explosions."

26. Amrine, "Cause of All the Rain?," 22.

27. Wexler, "Introductory Talk," Conference on Atomic Energy and Meteorology, U.S. Weather Bureau, December 19–20, 1949, Wexler Papers 16.

28. Wexler, "Statement for the Joint Congressional Committee on Atomic Energy . . . Regarding Nuclear Explosions and Weather," April 15, 1955, Wexler Papers 16; Wexler et al., "Atomic Energy and Meteorology."

29. Battle, "Documents on the U.S. Atomic Energy Detection System."

30. Machta, "Finding the Site of the First Soviet Nuclear Test"; U.S. Weather Bureau, "Report on Alert Number 112."

31. Machta to Berkner, October 6, 1961, Wexler Papers 13.

32. Machta, "Nuclear Explosions and Meteorology," 26.

33. Ibid., 26, 29.

34. Fleming, "Iowa Enters the Space Age."

35. Wexler, "Comments on Meteorological Control of Air Pollution."

36. Schrenk et al., "Air Pollution in Donora"; Snyder, "Death-Dealing Smog over Donora"; Davis, *When Smoke Ran Like Water*; Fleming and Knorr, "History of the Clean Air Act."

37. Wexler, "Comments on Meteorological Control of Air Pollution"; Wexler, "Great Smoke Pall."

38. Wexler, "Role of Meteorology in Air Pollution"; Brimblecombe, "Deciphering the Chemistry of Los Angeles Smog."

39. Wexler, "Meteorology and Air Pollution."

40. Ligda, "Radar Storm Observation," 1265.

41. Stratton, "Effect of Rain."

42. Rogers and Smith, "Short History."

43. Atlas and Ulbrich, "Early Foundations."

44. Guerlac, *RADAR in World War II*, 640; Buderi, *Invention That Changed the World*, 253.

45. Arthur E. Bent, Rad Lab, MIT, National Defense Research Committee to Reichelderfer, December 3, 1942, with cc. to Dr. L. A. Dubridge, classified SECRET, Reichelderfer Records, NARA, box 5.

46. Guerlac, *RADAR in World War II*, 640.

47. Ibid., 517.

48. Maynard, "Radar and Weather."

49. Ibid., 641–642.

50. Wexler, "Editorial Note to Accompany Commander Maynard's Paper," n.d. but ca. February 1946, Wexler Papers 2.

51. Wexler, "Structure of Hurricanes"; Atlas, *Radar in Meteorology*.

52. Maynard, "Radar and Weather"; Rogers and Smith, "Short History"; Buderi, *Invention That Changed the World*, 253.

53. Harry Wexler to Ray Wexler, September 11, 1946, Wexler Papers 2; R. Wexler, "Radar Detection of a Frontal Storm."

54. Duda, "A History of Radar Meteorology" (lyrics by Aaron Fleisher, melody unknown).

55. Rossby, "Current Problems in Meteorology," 36.

56. J. W. Mauchly, "Note on Possible Meteorological Use of High Speed Sorting and Computing Devices," April 14, 1945, marked "Confidential: copied from copy loaned by Prof. Mauchly 24 Jan. 1946," Wexler Papers 2.

57. Zworykin, "Outline of Weather Proposal."

58. Fleming, *Fixing the Sky*.

59. Reichelderfer to Zworykin, December 29, 1945, Wexler Papers 2; Harper, *Weather by the Numbers*.

60. Rossby to Reichelderfer, April 16, 1946, and reply May 9, 1946, Wexler Papers 2.

61. Wexler, "Notes from telephone conversation with C. G. Rossby," April 12, 1946; Rossby to Reichelderfer, April 16, 1946; Rossby to von Neumann, April 23, 1946; Wexler, "Trip Report, April 23–25, 1946," Wexler Papers 2.

62. F. Aydelotte, "Proposal to the U.S. Navy's Office of Research and Inventions," May 8, 1946, Wexler Papers 2; Wexler, "Trip Reports, May 7, and May 23–25, 1946," Wexler Papers 2; Thompson, "History of Numerical Weather Prediction."

63. Von Neumann to Wexler, June 28, 1946, Wexler Papers 2; Wexler to "Meteorology Computing Project," July 19, 1946, Wexler Papers 2.

64. Wexler, "Draft agenda, agenda, hand-written meeting notes, typed meeting notes, and trip report, Aug. 29–31, 1946," dated September 12, Wexler Papers 2; Nebeker, *Calculating the Weather*; Harper, *Weather by the Numbers*; Jaw, "Formation of the Semipermanent Centers."

65. Wexler, "Trip Report, November 13, 22, and December 7, 1946," Wexler Papers 2.

66. Participants from Chicago included John C. Bellamy, Werner Baum, George Benton, Jule Charney, George Cressman, David Fultz, Seymour Hess, Hsio-Lan Kuo, Alf Nyberg, Erik Palmén, George Platzman, Herbert Riehl, Carl-Gustaf Rossby, Zdenek Sekera, Victor Starr, Carl Christian Wallén, and Tu-Cheng Yeh. Attending from other institutions were Jerome Namias and Harry Wexler (U.S. Weather Bureau), Hans Panofsky (New York University), Paul Queney (Princeton University), K. Ryzkhov (USSR), Philip D. Thompson (U.S. Army), and Hurd Willett (MIT).

67. Wexler, "Notes on Problems of Meteorological Research," Chicago Conference, December 9–13, 1946, Wexler Papers 3.

68. Wexler, "Trip Report, October 30, 1946," Wexler Papers 2.

69. For details see Harper, *Weather by the Numbers*, 123ff.

70. Charney to Rossby, March 16, 1949; Rossby to Charney, March 27, 1949; Von Neumann to Rossby, June 13, 1949, Charney to Rossby, June 15, 1949, Charney Papers, MC 184, box 14, folder 459; Wexler to Reichelderfer, May 3, 1949, Wexler Papers 4.

71. Oppenheimer to Rossby, July 27, 1949, and September 28, 1949, and reply October 23, 1949, Institute for Advanced Study, Member Files, Rossby, B120.

72. Charney, Fjørtoft, and von Neumann, "Numerical Integration"; Charney to Platzman, April 10, 1950, in Platzman, "ENIAC Computations of 1950," 311–312; Platzman, "Some Remarks on High-Speed Automatic Computers."

73. Burks and Burks, "ENIAC," 387; Goldstine, *Computer*; Harper, *Weather by the Numbers*.

74. Phillips, "Interview."

75. Persson, "Early Operational Numerical Weather Prediction outside the USA . . . Part I."

76. J. Bjerknes to von Neumann, October 26, 1951, von Neumann Papers 15.

77. Cressman, "Origin and Rise."

78. "Practical Numerical Weather Prediction," meeting minutes, August 5, 1952, Wexler Papers 32.

79. Joint Chiefs of Staff, Joint Meteorological Committee, Ad Hoc Committee on Numerical Weather Prediction, Minutes, November 29, 1954, Wexler Papers 32; Thompson, "Interview"; J. Smagorinsky, "Beginnings of Numerical Weather Prediction"; "Forecaster Now Using Calculator."

80. Charney, "Use of the Primitive Equations"; "Man's Milieu," 76.

81. P. A. Sheppard to Von Neumann, June 15, 1954, Von Neumann Papers 15; Persson, "Early Operational Numerical Weather Prediction outside the USA . . . Part II," 269; Fjørtoft, "On a Numerical Method."

82. Phillips, "General Circulation"; Lewis, "Clarifying the Dynamics"; Anders Persson, personal communication to the author, July 22, 2014.

83. Rossby, "Current Problems in Meteorology," 36.

84. Pfeffer, *Dynamics of Climate*; Edwards, *Vast Machine*.

85. "Proposal for a Project on the Dynamics of the General Circulation," August 1, 1955, Von Neumann Papers 15; J. Smagorinsky, "Beginnings of Numerical Weather Prediction," 33.

86. M. Smagorinsky, "Interview."

87. Wexler memos to Al Carlin and to Helen, December 15, 1961, Wexler Papers 32.

88. This section is based on Fleming, "Earth Observations from Space."

89. Reichelderfer to James Van Allen, December 10, 1946, Wexler Papers 3; Reichhardt, "First Photo from Space"; Hubert and Berg, "Rocket Portrait of a Tropical Storm."

90. DeVorkin, *Science with a Vengeance*. See also Sheppard, "Meteorology"; Seaton, "Ionosphere"; Mitra, "General Aspects of Upper Atmospheric Physics."

91. H. Friedman, "The Pre-IGY Rocket Program."

92. William Kellogg, "Preliminary Proposal for Research on the Structure of the Stratosphere and Ionosphere," undated but ca. 1947, Wexler Papers 3.

93. Greenfield and Kellogg, "Inquiry into the Feasibility," v; J. Bjerknes, "Detailed Analysis of Synoptic Weather."

94. "Notes on Upper Atmospheric Nomenclature," Wexler Papers 32.

95. Committee on Extension to Standard Atmosphere to Wexler, May 3, 1962, Wexler Papers 14.

96. Wexler, "Antarctic Diary," Wexler Papers 27: Joseph Kaplan (1902–1991), chair of the U.S. National Committee (USNC) for the IGY; Laurence McKinley "Larry" Gould (1896–1995), director of the IGY Antarctic program; Lloyd V. Berkner (1905–1967), one of the architects of the IGY; A. Lincoln Washburn (1911–2007), USNC member; J. Wally Joyce (dates unknown), head of the IGY office at NSF; Hugh Odishaw (1916–1984), executive director of the USNC.

97. Wexler to Reichelderfer, June 24, 1955, Wexler Papers 7.

98. Wexler to Joseph Kaplan, November 16, 1955, Wexler Papers 7.

99. Wexler, "Antarctic Diary," August 15 and 20, 1956, Wexler Papers 27.

100. Ibid., October 18 and 20, 1956, Wexler Papers 27.

101. Wexler to Reichelderfer, December 19, 1956, Wexler Papers 8.

102. Wexler, "Antarctic Diary," August 25, 1956, Wexler Papers 27.

103. Wexler, "Meteorology in the International Geophysical Year."

104. Fleming and Seitchek, "Advancing Polar Research."

105. Wexler, "Meteorology in the International Geophysical Year."

106. Fleming, *Fixing the Sky*.

107. Reichelderfer to Wexler, February 25, 1949, Wexler Papers 4; Ben Holzman to Reichelderfer, March 1, 1949, Wexler Papers 4.

108. Wexler to Editor in Chief of *Time* Magazine, September 1, 1950, Wexler Papers 5; Wexler to C. F. Brooks, September 27, 1950, Wexler Papers 5.

109. Reichelderfer to C. F. Brooks, September 29, 1950, Wexler Papers 5.

110. US National Science Foundation, *Weather and Climate Modification*, 2–3; American Meteorological Society, *Statement on Weather Modification*, appendix C, Council Letter, March 8, 1957, Wexler Papers 9; Fleming, *Fixing the Sky*.

111. Reichelderfer to L. S. Rothschild, November 22, 1955, Wexler Papers 7; P. D. McTaggart-Cowan to C. F. Brooks, June 10, 1954, Wexler Papers 7.

112. Wexler, "Weather Modification and Space Exploration," Speech to National Rivers and Harbors Congress, Washington, DC, May 15, 1959, Wexler Papers 10.

113. President's Science Advisory Committee. *Introduction to Outer Space*, 1–2.

114. Wexler's handwritten speech is in Wexler Papers 16.

115. Fleming, "1954 Color Painting of Weather Systems," presents this image in full color.

116. Wexler, "Observing the Weather from a Satellite Vehicle"; Wexler, "Satellite and Meteorology"; Logsdon, "Meteorological Satellites."

117. President's Science Advisory Committee, *Introduction to Outer Space*.

118. Van Allen et al., "Observation of High Intensity Radiation by Satellites"; Van Allen and Frank, "Radiation around the Earth"; O'Keefe, Eckels, and Squires, "Pear-Shaped Component of the Geoid."

119. Fleming, "Iowa Enters the Space Age"; Fleming, "Earth Observations from Space"; Conway, *Atmospheric Science at NASA*.

120. Wexler to Weather Bureau Staff, September 10, 1958, Wexler Papers 9.

121. NASA Press Conference, "TIROS Television Infrared Observation Satellite," April 1, 1960, Wexler Papers 36,

122. NASA Press Conference, "TIROS I," April 22, 1960, Wexler Papers 36. Analysis of the early photos appears in Fritz and Wexler, "Cloud Pictures from Satellite"; Widger, "Examples of Project TIROS Data."

123. NASA Science Missions, http://science.nasa.gov/missions/tiros.

124. Purdom and Menzel, "Evolution of Satellite Observations."

125. Wexler, "Modifying Weather on a Large Scale."

126. Wexler, "On the Possibilities of Climate Control," 1962, Wexler Papers 18.

127. Fleming, *Callendar Effect.*

128. Wexler, "On the Possibilities of Climate Control," 1962, Wexler Papers 18.

129. Von Neumann, "Can We Survive Technology?"; Fleming, *Fixing the Sky.*

130. Wexler, "On the Possibilities of Climate Control," 1962, Wexler Papers 18.

131. Fleming, *Fixing the Sky,* 217.

132. Wexler, "On the Possibilities of Climate Control," 1962, Wexler Papers 18; Fleming, "Iowa Enters the Space Age."

133. Chapman, "Gases of the Atmosphere," 133.

134. Manabe and Möller, "On the Radiative Equilibrium"; Wexler, "On the Possibilities of Climate Control," 1962, Wexler Papers 18.

135. C. N. Touart to Wexler, January 11, 1962, and reply January 19, 1962, Wexler Papers 13.

136. Wexler, "Global Meteorology and the United Nations," 147.

137. President Kennedy's Inaugural Address, January 20, 1961, jfklibrary,org; State of the Union message, reading copy, January 30, 1961, jfklibrary.org; Dave Johnson to Wexler, Memo of Call, Wexler Papers 23.

138. "Possibilities in Meteorology for International Cooperation in Science," February 21, 1961, Wexler Papers 23.

139. John F. Kennedy News Conference no. 5, March 1, 1961, jfklibrary.org; John F. Kennedy Speeches. Excerpt, Address before a Joint Session of Congress, May 25, 1961, jfklibrary.org; John F. Kennedy Speeches. Address before the General Assembly of the United Nations, September 25, 1961, jfklibrary.org.

140. UN Resolution 1721; Malone, "International Cooperation in Meteorology."

141. Kennedy to Khrushchev, March 7, 1962, and reply, March 20, 1962, in Logsdon, *Exploring the Unknown*, 148 and 151.

142. Wexler, "Further Justification for the General Circulation Research Request for FY 63," February 9, 1962, Wexler Papers 18.

143. Wexler, "Chronology of Coordination Effected after Passage of U.N. Resolution 1721," For the Record Memo, July 23, 1962, Wexler Papers 14.

144. Wexler, "Global Meteorology and the United Nations."

145. Wexler to Reichelderfer, February 15, 1962, Wexler Papers 14.

146. Susan Wexler Schneider to the author, August 7, 2007, personal correspondence.

Chapter 5: Atmospheric Science

1. Petterssen, "The Atmospheric Sciences," 1962.

2. Hallgren, *University Corporation for Atmospheric Research*; Mazuzan, "Up, Up, and Away."

3. Rothschild to Bronk, September 27, 1955; Bronk to Byers et al., December 15, 1955, Committee on Meteorology, NAS Archives.

4. Reichelderfer to Bronk, March 14, 1956; Advisory Committee on Meteorology agenda and notes, April 3, 1956, Committee on Meteorology, NAS Archives.

5. Berkner was president of the Associated Universities, Inc. and the International Council of Scientific Unions and served as vice president of the International Scientific Radio Union, the American Geophysical Union, and the Special Committee for the International Geophysical Year. He was also a member of the President's Science Advisory Committee.

6. Minutes of second meeting, September 19, 1956, Committee on Meteorology, NAS Archives.

7. Third meeting report, November 27–28, 1956, Committee on Meteorology, NAS Archives.

8. Berkner to Bronk, April 5, 1957, Committee on Meteorology, NAS Archives. As of mid-June 1957, however, Charney had not yet formally accepted the appointment.

9. Berkner, "Horizons of Meteorology"; Hales, "Lloyd Viel Berkner," 3; Needell, *Science, Cold War and the American State*.

10. Berkner, "Horizons of Meteorology."

11. Ibid.

12. Ibid.

13. Launius, Fleming, and DeVorkin, *Globalizing Polar Science.*

14. Berkner, "Horizons of Meteorology."

15. Bascomb to Committee, Feb. 14, 1957, Committee on Meteorology, NAS Archives.

16. Ibid.

17. Berkner to Cornell, August 13, 1957, Committee on Meteorology, NAS Archives.

18. UCAR, *UCAR at 25,* 4.

19. Wexler, Antarctic Diary, January 17, 1958, Wexler Papers 27.

20. Chicago, Florida State, MIT, NYU, Penn State, UCLA, and Washington offered the PhD in meteorology; it was an option at Johns Hopkins, Oregon State, St. Louis, and Texas A&M.

21. U.S. National Academy of Sciences, Committee on Meteorology, *Research and Education in Meteorology.*

22. Mazuzan, "Up, Up, and Away."

23. Fleming, *Fixing the Sky,* 176–177.

24. U.S. National Academy of Sciences, Committee on Meteorology, "Research and Education in Meteorology"; Hallgren, *University Corporation for Atmospheric Research;* "Twelve Colleges Plan Weather Studies." Houghton was trained as an electrical engineer, not as a meteorologist, and did not have a doctorate, but he was chosen as chair of the meteorology department at MIT. Houghton, Curriculum Vitae and related documents, Henry Houghton Papers B1, F1.

25. Thomas Malone to Paul Klopsteg, July 17, 1958; Klopsteg to Detlev Bronk, July 29, 1958; S. D. Cornell to Klopsteg, August 16, 1958, Committee on Atmospheric Sciences, NAS Archives.

26. U.S. National Academy of Sciences, Committee on Atmospheric Sciences, *Proceedings of the Scientific Information Meeting on Atmospheric Sciences, November 24, 1958.*

27. University Committee on Atmospheric Research, *Preliminary Plans.*

28. Ibid., appendix E; Malone, "A National Institute."

29. Hallgren, *University Corporation for Atmospheric Research;* UCAR, *UCAR at Twenty-five,* 1. A photo of the Tucson meeting is at http://www.ucar.edu/communications/ucar25/experiment.html.

30. Fleming, "Iowa Enters the Space Age."

31. Bassi, *A Scientific Peak.*

32. Ibid., 148.

33. "Walter Orr Roberts Is Dead at Seventy-Four."

34. UCAR, *Remembering Walt Roberts,* 81–82, 86; Fleagle, *Eyewitness,* 86; Leslie, "A Different Kind of Beauty."

35. Platzman, *Conversations with Jule Charney,* 133–135.

36. President's Science Advisory Committee, *Introduction to Outer Space*; Harper, *Weather by the Numbers,* 176ff; Thompson, "Interview"; Fleagle, "From the International Geophysical Year to Global Change."

37. Petterssen, *Weathering the Storm.*

38. "Plan Weather Program"; U.S. National Academy of Sciences, Committee on Atmospheric Sciences, *The Atmospheric Sciences, 1961–1971.*

39. Reichelderfer, Memo on coordination in the atmospheric sciences, June 19, 1961, Reichelderfer Records, NARA, box 3; U.S. Senate, *Organizing for National Security.*

40. Reichelderfer to John F. Kennedy, July 24, 1963, Library of Congress, Reichelderfer Papers 7.

41. Namely, the Atomic Energy Commission, Bureau of the Budget, Federal Aviation Agency, Federal Communications Commission, National Academy of Sciences, National Aeronautics and Space Administration, and Office of Science and Technology, and departments of agriculture, commerce, defense, health, education, and welfare, interior, and state.

42. Sverre Petterssen, chair, "Report of the Ad-Hoc Committee on International Cooperation in the Atmospheric Sciences," September 1, 1961, Committee on Atmospheric Sciences, NAS Archives; U.S. National Academy of Sciences, Committee on Atmospheric Sciences, *The Atmospheric Sciences, 1961–1971,* 83.

43. UN Resolution 1721 (XVI); Malone, "International Cooperation in Meteorology."

44. Charney to Wexler and Multiple Recipients, April 5, 1962, Enclosing "A Suggested Meteorological Observation Plan," Wexler Papers 14.

45. Ibid.

46. Lorenz, "Statistical Prediction of Solutions of Dynamic Equations"; Lorenz, *Essence of Chaos*; Yoden, "Atmospheric Predictability"; Anders Persson, personal communication to the author, July 22, 2014.

Chapter 6: Final Thoughts

1. Bergeron, "Methods in Scientific Weather Analysis and Forecasting," 468.

2. Lorenz, "A History of Prevailing Ideas"; Lorenz, *Essence of Chaos*; Lorenz, "Evolution of Dynamic Meteorology."

3. Wexler, "Satellite and Meteorology," 8; Rossby, "Current Problems in Meteorology," 36.

8 Bibliography

Bjerknes Publications

"Publications by Vilhelm Bjerknes." *In Memory of Vilhelm Bjerknes on the Hundreth Anniversary of His Birth*, edited by A. Eliassen and E. Hoiland. Oslo: Universitetsforlaget. *Geofysiske Publikasjoner* 24 (1962): 26–37. http://www.colby.edu/sts/bjerknes.pdf.

Rossby Publications

Byers, Horace R. "Carl-Gustaf Arvid Rossby, 1898–1957." *Biographical Memoirs. National Academy of Sciences* (1960): 265–270. http://www.colby.edu/sts/rossby.pdf.

Wexler Publications

"Bibliography of the Publications of Harry Wexler." Compiled by M. Rigby and P. A. Keehn. *Monthly Weather Review* 91 (1963): 477–481. http://www.colby.edu/sts/wexler.pdf.

Archival Sources

Bergen Geophysical Institute, Bergen, Norway.

  Vervarslinga på Vestlandet Archives [VpV Archives].

Carnegie Institution for Science (CIS) Archives, Washington, DC.

  Advisory Committee on Meteorology.

  Vilhelm Bjerknes Files.

Institute for Advanced Study, Shelby White and Leon Levy Archives Center, Princeton, NJ.

Records of the Office of the Director, Member Files, Rossby, C.G.A., B120.

Kennedy, John F. Presidential Library and Museum, http://www.jfklibrary.org.

Kungliga Vetenskapsakademien Archives, Stockholm.

Fysiska Sällskapets i Stockholm, Protokollsbok 1 (1891–1898) and 2 (1899–1915).

Lewis, John M. Personal Research Collection on C.-G. Rossby.

Massachusetts Institute of Technology, Institute Archives and Special Collections, Cambridge, MA.

Jule G. Charney Papers, MC 184.

Henry Houghton Papers, MC 242.

MIT President's Reports, 1927–1940, http://libraries.mit.edu/archives/mithistory/presidents-reports.html.

Office of the President, AC 4 and AC 13.

University Meteorology Committee, MC 511.

Nasjonalbilbioteket, Oslo, Norway (NBO).

Bjerknes, Vilhelm, Brevs 469B.

Familien Bjerknes, Brevs 469C.

President's Science Advisory Committee, 1957–1961. Papers microfilmed from the holdings of the Dwight D. Eisenhower Library.

Statsarkivet i Bergen.

Textual and cartographic records related to Vervarslinga på Vestlandet.

Stockholm University.

Meteorologiska Institutionen, Professor Carl-Gustaf Rossbys Korrespondens, 1948–1957.

University of Chicago Library, Special Collections.

Department of Meteorology Records.

U.S. Library of Congress, Manuscript Division, Washington, DC.

Papers of Lloyd V. Berkner.

Papers of John von Neumann.

Papers of Francis Wilton Reichelderfer.

Papers of Harry Wexler.

U.S. National Academy of Sciences (NAS) Archives.

Committee on Atmospheric Sciences.

Committee on Meteorology.

U.S. National Archives and Records Administration (NARA).

Record Group 27, Records of the Weather Bureau, Records of the Office of the Chief, Office Files of F. W. Reichelderfer, General Correspondence, 1939–63 (Reichelderfer Records, NARA).

Record Group 331, Military Agency Records.

Wexler Family Papers. Susan Wexler Schneider.

References

Abbe, Cleveland. *The Mechanics of the Earth's Atmosphere: A Collection of Translations, Second Collection. Smithsonian Miscellaneous Collections 843*. Washington, DC: Smithsonian Institution, 1891.

Abbe, Cleveland. *The Mechanics of the Earth's Atmosphere: A Collection of Translations, Third Collection. Smithsonian Miscellaneous Collections 51, no. 4*. Washington, DC: Smithsonian Institution, 1910.

Abbe, Cleveland. "The Needs of Meteorology," *Science* 1, no. 7 (1895), 181–182.

Abbe, Cleveland. "The Physical Basis of Long-Range Weather Forecasts." *Monthly Weather Review* 29 (1901): 551–561.

Abbe, Cleveland. "The Weather Map on the Polar Projection." *Monthly Weather Review* 42 (1) (1914): 36–38.

Aitken, Hugh G. J. *The Continuous Wave: Technology and American Radio, 1900–1932*. Princeton, NJ: Princeton University Press, 1985.

Amrine, Michael. "Cause of All the Rain? Weather Is Unchanged by A-Bomb; Dwarfs It," *Boston Globe*, April 24, 1953, 22.

Appleton, Edward V. "The Ionosphere." Nobel Lecture, December 12, 1947. http://www.nobelprize.org/nobel_prizes/physics/laureates/1947/appleton-lecture.html.

Arrhenius, Svante. *Worlds in the Making: The Evolution of the Universe*. Translated by H. Borns. New York: Harper, 1908.

Ashford, Oliver M. *Prophet—or Professor? The Life and Work of Lewis Fry Richardson*. Bristol: A. Hilger, 1985.

Aspray, William. *John Von Neumann and the Origins of Modern Computing*. Cambridge, MA: MIT Press, 1990.

Atlas, D., and C. W. Ulbrich. "Early Foundations of the Measurement of Rainfall by Radar." In *Radar in Meteorology*, edited by D. Atlas, 56–97. Boston: American Meteorological Society, 1990.

Bailey, Floyd P. *Santa Rosa Junior College, 1918–1957: A Personal History*. Santa Rosa, CA: Santa Rosa Junior College, 1967.

Bassi, Joseph P. *A Scientific Peak: How Boulder Became a World Center for Space and Atmospheric Science*. Boston: AMS Books, 2015.

Bates, Charles C. "The Formative Rossby-Reichelderfer Period in American Meteorology, 1926–40." *Weather and Forecasting* 4 (1989): 593–603.

Bates, Charles C., and John F. Fuller. *America's Weather Warriors, 1814–1985*. College Station: Texas A&M University Press, 1986.

Battle, Joyce. "Documents on the U.S. Atomic Energy Detection System [AEDS]." National Security Archive Electronic Briefing Book No. 7. http://www2.gwu.edu/~nsarchiv/NSAEBB/NSAEBB7/nsaebb7.htm.

Beck, Anne Louise. "An Application of the Principles of Bjerknes' Dynamic Meteorology in a Study of Synoptic Weather Maps for the United States." MA thesis, University of California at Berkeley, 1922.

Beck, Anne Louise. "The Earth's Atmosphere as a Circular Vortex." *Monthly Weather Review* 50 (1922): 393–401.

Bellamy, John C. *Prospectus of Meteorological Operations. U.S. Weather Bureau Technical Planning Study No. 2*. Washington, DC: U.S. Department of Commerce, 1960.

"The Bergen Geophysical Institute: 'Polar-Front' Weather Forecasting." *Bulletin of the American Meteorological Society* 2 (1921): 51–52.

Bergeron, Tor. "Methods in Scientific Weather Analysis and Forecasting: An Outline in the History of Ideas and Hints at a Program." In *The Atmosphere and the Sea in Motion: Scientific Contributions to the Rossby Memorial Volume*, edited by B. Bolin, 440–474. New York: Rockefeller Institute Press and Oxford University Press, 1959.

Bergeron, Tor. "The Young Carl-Gustaf Rossby." In *The Atmosphere and the Sea in Motion: Scientific Contributions to the Rossby Memorial Volume*, edited by B. Bolin, 51–55. New York: Rockefeller Institute Press and Oxford University Press, 1959.

Bergeron, Tor, Olaf Devik, and Carl Ludvig Godske. "Vilhelm Bjerknes, March 14, 1862–April 9, 1951." In *In Memory of Vilhelm Bjerknes on the Hundreth Anniversary of His Birth*, edited by A. Eliassen and E. Hoiland, *Geofysiske Publikasjoner* 24, 7–25. Oslo: Universitetsforlaget, 1962.

Berkner, Lloyd. "Horizons of Meteorology." Typescript, May 1, 1957, Library of Congress, Lloyd Berkner Papers, Box 12; reprinted in *Bulletin of the American Meteorological Society* 81 (2000): 2969–2973.

Berry, F. A., E. Bollay, and N. R. Beers, eds. *Handbook of Meteorology*. New York: McGraw-Hill, 1945.

Bjerknes, Carl Anton. "Hydrodynamiske Analogier til de statisk elektriske og de magnetiske Kräfter." *Naturen*. Kristiania: Mallingske Bogtrykkeri, 1880.

Bjerknes, Jacob. "Detailed Analysis of Synoptic Weather as Observed from Photographs Taken on Two Rocket Flights over White Sands, New Mexico, July 26, 1948." Appendix I in S. M. Greenfield and W. W. Kellogg, U.S. Air Force, Project RAND, "Inquiry into the Feasibility of Weather Reconnaissance from a Satellite Vehicle," Secret report, April. Santa Monica, CA: Rand, 1951.

Bjerknes, Jacob. "Half a Century of Change in the 'Meteorological Scene.'" *History of Meteorology* 6 (2014): 1–6.

Bjerknes, Jacob. "On the Structure of Moving Cyclones." *Geofysiske Publikationer* 1 (2) (1919); *Monthly Weather Review* 47 (2) (1919): 95–99.

Bjerknes, Jacob. "Practical Examples of Polar-Front Analysis over the British Isles in 1925–6." Air Ministry Publication 307. London: HMSO, 1930.

Bjerknes, Jacob. "Über die Fortbewegung der Konvergenz—und Divergenzlinien." *Meteorologische Zeitschrift* 34 (1917): 345–349.

Bjerknes, Jacob, and M. A. Giblett. "An Analysis of a Retrograde Depression in the Eastern United States of America." *Monthly Weather Review* 52 (1924): 521–527.

Bjerknes, Jacob, and H. Solberg. "Life Cycle of Cyclones and the Polar Front Theory of Atmospheric Circulation." *Geofysiske Publikasjoner* 3 (1) (1922): 3–18.

Bjerknes, Jacob, and H. Solberg. "Meteorological Conditions for the Formation of Rain." *Geofysiske Publikasjoner* 2 (3) (1921): 60p.

Bjerknes, Vilhelm. *C. A. Bjerknes: Sein leben und seine arbeit*. Translated by E. W. Köppen. Berlin: Julius Springer, 1933.

Bjerknes, Vilhelm. "The Circulatory Movements in the Atmosphere." *Monthly Weather Review* 28 (12) (1900): 532–535.

Bjerknes, Vilhelm. "Cirkulation relativ zu der Erde." *Meteorologische Zeitschrift* 19 (1902): 97–108.

Bjerknes, Vilhelm. "A Course of Twenty-nine Lectures on Physical Hydrodynamics." Delivered in the Norman Bridge Laboratory of Physics at the California Institute of Technology, Pasadena, California during October, November, and December, 1924. Mimeo, 1925.

Bjerknes, Vilhelm. "Dynamics and Electromagnetism." *Avhandlinger utgitt av Det norske videnskaps-akademi i Oslo*, Mat.-naturv. kl. no. 1 (1949): 12 pp.

Bjerknes, Vilhelm. "The Dynamic Principles of the Circulatory Movements in the Atmosphere." *Monthly Weather Review* 28 (10) (1900): 434–443.

Bjerknes, Vilhelm. "Das dynamische Princip der Cirkulationsbewegungen in der Atmosphäre." *Meteorologische Zeitschrift* 17 (1900): 97–106.

Bjerknes, Vilhelm. *Fields of Force: A Course of Lectures in Mathematical Physics Delivered December 1 to 23, 1905.* New York: Columbia University Press, 1906.

Bjerknes, V. "The Forces Which Lift Aeroplanes." *Nature* 114 (1924): 472–474, 508–510.

Bjerknes, Vilhelm. "Geofysiker motet i Göteborg den 28–31 augusti 1918: forhandlingar." Utgivna av Otto Nordenskjold och Hans Pettersson. Göteborg: Göteborgs handelstidnings aktiebolags tryckeri, 1919.

Bjerknes, Vilhelm. "Hvordan Bergensskolen ble til." *Vervarslinga på Vestlandet 25 år Festskrift*, 7–19. Bergen: A. S. John Griegs Boktrykkeri, 1944.

Bjerknes, Vilhelm. "Leipzig-Bergen. Festvortrag zur 25-Jahrfeier des Geophysikalischen Instituts der Universität Leipzig." *Zeitschrift für Geophysik* 14 (1938): 49–62.

Bjerknes, Vilhelm. "Die Meteorologie als Exacte Wissenshaft." In *Antritsvorlesung Gehalten am 8 Januar 1913 in der Aula der Universität Leipzig.* Braunschweig: Vieweg & Sohn, 1913.

Bjerknes, Vilhelm. "Meteorology as an Exact Science." *Monthly Weather Review* 42 (1) (1914): 11–14.

Bjerknes, Vilhelm. "The Meteorology of the Temperate Zone and the General Atmosphere Circulation." *Nature* 105 (1920): 522–524.

Bjerknes, Vilhelm. "Nyere hydrodynamiske Undersøgelser." *Naturen* (Bergen) 6 (1882): 132–146.

Bjerknes, Vilhelm. "Om elektricitetsbevregelsen i Hertz's primrere leder." *Archiv for Mathematik og Naturvidenskab* 15 (1892): 165–236.

Bjerknes, Vilhelm. "On the Dynamics of the Circular Vortex with Applications to the Atmosphere and Atmospheric Vortex and Wave Motions." *Geofysiske Publikasjoner* 2 (4) (1921): 88p.

Bjerknes, Vilhelm. "Possible Improvements in Weather Forecasting with Special Reference to the United States." *Monthly Weather Review* 47 (2) (1919): 99–100.

Bjerknes, Vilhelm. "Das Problem der Wettervorhersage, betrachtet vom Standpunkte der Mechanik und der Physik." *Meteorologische Zeitschrift* 21 (1904): 1–7.

Bjerknes, Vilhelm. "Solar Hydrodynamics." *Astrophysical Journal* 64 (1926): 93–121.

Bjerknes, Vilhelm. "Synoptical Representation of Atmospheric Motions." *Quarterly Journal of the Royal Meteorological Society* 36 (1910): 267–286.

Bjerknes, Vilhelm. "Über die Bildung von Cirkulationsbewegungen und Wirblen in reibungslosen Flüssigkeiten." *Videnskabsselskabets Skrifter* 1, *Mathematisknaturvidenskabelig Klasse*, no. 5. Kristiania: Mallingske Bogtrykkeri, 1898.

Bjerknes, Vilhelm. "Über einen hydrodynamischen Fundametalsatz und seine Anwendung besonders auf die Mechanik der Atmosphäre und des Weltmeeres." *Kongliga Svenska Vetenskaps-Akademiens Handlingar* 31 (4) (1898): 35 pp.

Bjerknes, Vilhelm. *Vorlesungen über hydrodynamische Fernkräfte nach C. A. Bjerknes's Theorie*, 2 vols. Leipzig: Barth, 1900–1902.

Bjerknes, Vilhelm, J. Bjerknes, H. Solberg, and T. Bergeron. *Hydrodynamique physique*. Paris: Presses Universitaires de France, 1934.

Bjerknes, Vilhelm, J. Bjerknes, H. Solberg, and T. Bergeron. *Physikalische Hydrodynamik mit Anwendung auf die dynamische Meteorologie*. Berlin: Springer, 1933.

Bjerknes, Vilhelm, and collaborators. *Dynamic Meteorology and Hydrography*. Part 2, *Kinematics*. Washington, DC: Carnegie Institution of Washington, 1911.

Bjerknes, Vilhelm, and J. W. Sandström. *Dynamic Meteorology and Hydrography*, Part I, *Statics*. Washington, DC: Carnegie Institution of Washington, 1910.

Bohn, Maria. "Concentrating on CO_2: The Scandinavian and Arctic Measurements." *Osiris* 26 *Klima* (2011): 165–179.

Bolin, Bert, ed. *The Atmosphere and the Sea in Motion: Scientific Contributions to the Rossby Memorial Volume*. Stockholm: Rockefeller Institute Press and Oxford University Press, 1959.

Bolin, Bert. "Carl-Gustaf Rossby in Memoriam." *Tellus* 9 (1957): 257–258.

Bolin, Bert. "Carl-Gustaf Rossby: The Stockholm Period, 1947–1957." *Tellus. Series B, Chemical and Physical Meteorology* 51 (1999): 4–12.

Bolin, Bert, and Erik Eriksson. "Changes in the Carbon Dioxide Content of the Atmosphere and Sea Due to Fossil Fuel Combustion." In *The Atmosphere and the Sea in Motion: Scientific Contributions to the Rossby Memorial Volume*, edited by B. Bolin, 130–142. New York: Rockefeller Institute Press, 1959.

Bonacina, Leo W. C. "The Great Problem of Meteorology." *Symons's Meteorological Magazine* 40 (1905): 7–10.

Bowie, Edward Hall. *Weather and the Airplane: A Study of the Model Weather Reporting Service over the California Airway*. New York: Guggenheim, 1929.

Braham, Roscoe R., Jr., and Thomas F. Malone. "Horace Robert Byers, 1906–1998." *Biographical Memoirs*, vol. 79, 3–19. Washington, DC: National Academy of Sciences, 2001.

Brimblecombe, Peter. "Deciphering the Chemistry of Los Angeles Smog, 1945–95." In *Toxic Airs: Body, Place, Planet in Historical Perspective*, edited by J. R. Fleming and A. Johnson, 95–108. Pittsburgh: University of Pittsburgh Press, 2014.

"British Association Meeting at Toronto, August 1924." *Quarterly Journal of the Royal Meteorological Society* 50 (1924): 324.

Brooks, Charles F. "Free Air Conditions." *Geographical Review* 9 (4) (1920): 372.

Brown, Louis. *Centennial History of the Carnegie Institution of Washington*. Vol. 2, *The Department of Terrestrial Magnetism*. Cambridge: Cambridge University Press, 2005.

Buderi, Robert. *The Invention That Changed the World: How a Small Group of Radar Pioneers Won the Second World War and Launched a Technological Revolution*. New York: Simon & Schuster, 1996.

Bumstead, H. A. "Atmospheric Radio-activity." *American Journal of Science*, 4th ser. 18 (1904): 1–11.

Burks, Arthur W., and Alice Burks. "The ENIAC: First General-Purpose Electronic Computer." *Annals of the History of Computing* 3 (1981): 310–389.

Byers, Horace R. "Carl-Gustaf Rossby, the Organizer." In *The Atmosphere and the Sea in Motion: Scientific Contributions to the Rossby Memorial Volume*, edited by B. Bolin, 56–59. New York: Rockefeller Institute Press and Oxford University Press, 1959.

Byers, Horace R. "The Founding of the Institute of Meteorology at the University of Chicago." *Bulletin of the American Meteorological Society* 57 (1976): 1343–1345.

Byers, Horace R. "Recollections of the War Years." *Bulletin of the American Meteorological Society* 51 (1970): 214–217.

Carlson, Per. "The Discovery of Cosmic Rays." Webinar at the American Physical Society, May 1, 2011. http://apps3.aps.org/aps/meetings/april11/presentations/h13 -1-carlson.pdf.

Carnegie, Andrew. "Trust Deed." *The Carnegie Institution of Washington, D.C., Founded by Andrew Carnegie*. Washington, DC: Carnegie Institution, 1902.

Carnegie Institution Year Books, 1–51 (1902–1951). http://carnegiescience.edu/ publications/yearbooks.

Cartwright, Gordon D., and Charles H. Sprinkle. "A History of Aeronautical Meteorology: Personal Perspectives, 1903–1995." In *Historical Essays on Meteorology, 1919–1995*, edited by J. R. Fleming, 443–480. Boston: American Meteorological Society, 1996.

Chapman, Sydney. "The Gases of the Atmosphere." *Quarterly Journal of the Royal Meteorological Society* 60 (1934): 127–142.

Charney, Jule G. "The Dynamics of Long Waves in a Baroclinic Westerly Current." *Journal of Meteorology* 4 (1947): 135–162.

Charney, Jule G. "On a Physical Basis for Numerical Prediction of Large-Scale Motions in the Atmosphere." *Journal of Meteorology* 6 (1949): 371–385.

Charney, Jule G. "Radiation." In *Handbook of Meteorology*, edited by F. A. Berry, E. Bollay, and N. R. Beers, 284–311. New York: McGraw-Hill, 1945.

Charney, Jule G. "The Use of the Primitive Equations of Motion in Numerical Prediction." *Tellus* 7 (1955): 22–26.

Charney, Jule G., R. Fjørtoft, and J. von Neumann. "Numerical Integration of the Barotropic Vorticity Equation." *Tellus* 2 (1950): 237–254.

Christie, J.R.R. "Aurora, Nemesis and Clio." *British Journal for the History of Science* 26 (1993): 391–405.

Coen, Deborah R. "Scaling Down: The 'Austrian' Climate between Empire and Republic." In *Intimate Universality: Local and Global Themes in the History of Weather and Climate*, edited by J. R. Fleming, V. Jankovic, and D. R. Coen, 115–140. Sagamore Beach, MA: Science History Publications, 2006.

Coen, Deborah R. "A Scientific Dynasty: Probability, Liberalism, and the Exner Family in Imperial Austria." PhD dissertation, Harvard University, 2004.

Coen, Deborah R. 2008. *Vienna in the Age of Uncertainty: Science, Liberalism, and Private Life*. Chicago: University of Chicago Press.

"A Conference in Bergen." *Meteorological Magazine* (September 1920): 166–168.

Conway, Erik M. *Atmospheric Science at NASA: A History*. Baltimore, MD: Johns Hopkins University Press, 2008.

Cressman, George. "The Origin and Rise of Numerical Weather Prediction." In *Historical Essays on Meteorology, 1919–1995*, ed. J. R. Fleming, 21–39. Boston: American Meteorological Society, 1996.

Curie, Pierre. "Radioactive Substances, Especially Radium." Nobel Lecture, June 6, 1905. http://www.nobelprize.org/nobel_prizes/physics/laureates/1903/pierre-curie-lecture.html.

Dadourian, H. M. "On the Constituents of Atmospheric Radioactivity." *American Journal of Science* 4th ser. 25 (1908): 335–342.

Darrigol, Oliver. *Worlds of Flow: A History of Hydrodynamics from Bernoulli to Prandtl*. Oxford: Oxford University Press, 2005.

Davies, H. C. "Emergence of the Mainstream Cyclogenesis Theories." *Meteorologische Zeitschrift* n.f. 6 (1997): 261–274.

Davis, Devra Lee. *When Smoke Ran Like Water: Tales of Environmental Deception and the Battle against Pollution*. New York: Basic Books, 2002.

Devereaux, W. C. "A Meteorological Service of the Future." *Bulletin of the American Meteorological Society* 29 (1939): 212–221.

DeVorkin, David H. *Science with a Vengeance: How the Military Created the U.S. Space Sciences after World War II*. New York: Springer-Verlag, 1992.

Douglas, Susan J. *Inventing American Broadcasting, 1899–1922*. Baltimore, MD: Johns Hopkins University Press, 1987.

Doyle, J. D., C. C. Epifanio, A. Persson, P. A. Reinecke, and G. Zängl. "Mesoscale Modeling over Complex Terrain: Numerical and Predictability Perspectives." In *Mountain Weather Research and Forecasting: Recent Progress and Current Challenges*, edited by F. K. Chow, S.F.J. De Wekker, and B. J. Snyder, 531–589. Dordrecht: Springer, 2013.

DuBois, John L., Robert P. Multhauf, and Charles A. Ziegler. *The Invention and Development of the Radiosonde, with a Catalog of Upper-Atmospheric Telemetering Probes in the National Museum of American History, Smithsonian Institution*. Smithsonian Studies in History and Technology, no. 53. Washington, DC: Smithsonian Institution Press, 2002.

Duda, Jeff. "A History of Radar Meteorology: People, Technology, and Theory." Lecture notes, http://www.meteor.iastate.edu/~jdduda/portfolio/HistoryPPT.pdf.

Durfee High School. *Record*. Fall River, MA, 1928.

Edwards, Paul N. *A Vast Machine: Computer Models, Climate Data, and the Politics of Global Warming*. Cambridge, MA: MIT Press, 2010.

Ekholm, Nils. *Étude des conditions météorologiques à l'aide de cartes synoptiques: représentant la densité de l'air*. Stockholm: P. A. Norstedt, 1891.

Eliassen, Arnt. "Jacob Aall Bonnevie Bjerknes (1897–1975)." In *Biographical Memoirs*, vol. 58, 3–21. Washington, DC: National Academy of Sciences, 1995.

Ellingsen, Gunnar. "Varme havstrømmer og kald krig: 'Bergensstrømmåleren' og vitenskapen om havstrømmer fra 1870-årene til 1960-årene." PhD diss., Universitetet i Bergen, 2012.

Elsaesser, W. M. *Heat Transfer by Infrared Radiation in the Atmosphere*. Harvard Meteorological Studies no. 6. Milton, MA: Blue Hill Meteorological Observatory, 1942.

Eriksson, E. "Report on an Informal Conference in Atmospheric Chemistry Held at the Meteorological Institute, University of Stockholm, May 24–26, 1954." *Tellus* 4 (1954): 302–307.

Eriksson, E. "Report on the Second Informal Conference on Atmospheric Chemistry, Held at the Meteorological Institute, University of Stockholm, May 31–June 4, 1955." *Tellus* 5 (1955): 388–394.

Euler, Leonhard. *Letters on Different Subjects in Natural Philosophy: Addressed to a German Princess*. Vol. 1. New York, 1837.

Eve, Arthur S. "On the Ionization of the Atmosphere Due to Radioactive Matter." *Philosophical Magazine* 6th ser. 21 (1911): 26–40.

Exner, Felix M. "Anschauungen über kalte und warme Luftströmungen nahe der Erdoberfläche und ihre Rolle in den niedrigen Zyklonen. (Vortrag, Gehalten auf der Meteorologenkonferenz in Bergen im August 1920)." *Geografiska Annaler* 2 (1920): 225–236.

Exner, Felix M. *Dynamische Meteorologie*. Leipzig: Teubner, 1917; 2nd ed. Vienna: Julius Springer, 1925.

"Fellowship in Meteorology." *Bulletin of the American Meteorological Society* 1 (1920): 95–96.

Ficker, Heinrich von. "Polarfront, Aufbau, Entstehung und Lebensgeschiehte der Zyklonen." *Meteorologische Zeitschrift* 40 (3) (1923): 65–79, 264–267.

Fjørtoft, R. "On a Numerical Method of Integrating the Barotropic Vorticity Equation." *Tellus* 4 (1952): 179–194.

Fleagle, Robert G. *Eyewitness: Evolution of the Atmospheric Sciences*. Boston: American Meteorological Society, 2001.

Fleagle, Robert G. "From the International Geophysical Year to Global Change." *Reviews of Geophysics* 30 (1992): 305–313.

Fleming, J. R. "Beyond Prediction to Climate Modeling and Climate Control: New Perspectives from the Papers of Harry Wexler, 1945–1962." In *The Development of Atmospheric General Circulation Models*, edited by Leo Donner, Wayne Schubert, and Richard Somerville, 51–75. Cambridge: Cambridge University Press, 2011.

Fleming, J. R. *The Callendar Effect*. Boston: AMS Books, 2007.

Fleming, J. R., ed. *Climate Change and Anthropogenic Greenhouse Warming: A Selection of Key Articles, 1824–1995, with Interpretive Essays*. Classic Articles in Context. National Science Digital Library, 2008. http://bit.ly/15CTWnT.

Fleming, J. R. "Earth Observations from Space: The First Two Decades." In *Earth Observations from Space: The First Fifty Years of Scientific Achievements*, edited by Committee on Scientific Accomplishments of Earth Observations from Space, National Research Council, 14–27. Washington, DC: National Academies Press, 2008.

Fleming, J. R. *Fixing the Sky: The Checkered History of Weather and Climate Control*. New York: Columbia University Press, 2010.

Fleming, J. R., ed. *Historical Essays on Meteorology, 1919–1995*. Boston: American Meteorological Society, 1996.

Fleming, J. R. *Historical Perspectives on Climate Change*. New York: Oxford University Press, 1998.

Fleming, J. R. "Iowa Enters the Space Age: James Van Allen, Earth's Radiation Belts, and Experiments to Disrupt Them." *Annals of Iowa* 70 (2011): 301–324.

Fleming, J. R. *Meteorology in America, 1800–1870*. Baltimore, MD: Johns Hopkins University Press, 1990.

Fleming, J. R. "A 1954 Color Painting of Weather Systems as Viewed from a Future Satellite." *Bulletin of the American Meteorological Society* 88 (2007): 1525–1527. http://bit.ly/406earthfromspace1954.

Fleming, J. R. "Polar and Global Meteorology in the Career of Harry Wexler, 1933–1962." In *Globalizing Polar Science: Reconsidering the Social and Intellectual Implications of the International Polar and Geophysical Years*, edited by R. D. Launius, J. R. Fleming, and D. H. DeVorkin, 225–241. New York: Palgrave, 2010.

Fleming, J. R. "Sverre Petterssen and the Contentious (and Momentous) Weather Forecasts for D-Day." *Endeavour* 28 (June 2004): 59–63.

Fleming, J. R., and Bethany Knorr. "History of the Clean Air Act." 1997. http://www.ametsoc.org/sloan/cleanair/index.html.

Fleming, J. R., and Cara Seitchek. "Advancing Polar Research and Communicating Its Wonders: Quests, Questions, and Capabilities of Weather and Climate Studies in International Polar Years." In *Smithsonian at the Poles: Contributions to International Polar Year Science*, edited by I. Krupnik, M. A. Lang, and S. E. Miller, 1–12. Washington, DC: Smithsonian Institution Scholarly Press, 2009.

Flight and Aircraft Engineer 38 (1940) and 61 (1952).

Fonselius, S., and F. Koroleff, with a preface by K. Buch. "Microdetermination of CO_2 in the Air, with Current Data for Scandinavia." *Tellus* 7 (1955): 258–265.

Forbes, George. "Hydrodynamic Analogies to Electricity and Magnetism." *Nature* 24 (1881): 360–361.

"Forecaster Now Using Calculator," *Washington Post and Times Herald*, May 8, 1955, A12.

Friedman, Herbert. "The Pre-IGY Rocket Program of the United States." In *Geophysics and the IGY*, edited by H. Odishaw and S. Ruttenberg, 108–118. Washington, DC: American Geophysical Union of the National Academy of Sciences, National Research Council, 1958.

Friedman, Robert Marc. *Appropriating the Weather: Vilhelm Bjerknes and the Construction of a Modern Meteorology*. Ithaca, NY: Cornell University Press, 1989.

Friedman, Robert Marc. "Bjerknes, Vilhelm." In *Complete Dictionary of Scientific Biography*, vol. 19, 288–290. Detroit: Charles Scribner's Sons, 2008.

Fritz, Sigmund, and Harry Wexler. "Cloud Pictures from Satellite TIROS I." *Monthly Weather Review* 88 (1960): 79–87.

Godske, C. L., T. Bergeron, J. Bjerknes, and R. C. Bundgaard. *Dynamic Meteorology and Weather Forecasting*. Boston: American Meteorological Society and Washington, DC: Carnegie Institution, 1957.

Gold, E. "Fronts and Occlusions." *Quarterly Journal of the Royal Meteorological Society* 61 (1935): 107–157.

Gold, E. "Vilhelm Friman Koren Bjerknes, 1862–1951." *Obituary Notices of Fellows of the Royal Society* 7 (20) (1951): 302–317.

Goldstine, Herman H. *The Computer from Pascal to von Neumann*. Princeton, NJ: Princeton University Press, 1972.

Greenfield, S. M., and W. W. Kellogg, U.S. Air Force, and Project RAND, "Inquiry into the Feasibility of Weather Reconnaissance from a Satellite Vehicle," Secret report, April. Santa Monica, CA: Rand, 1951.

Gregg, Willis R. "Meteorological Service for Airways in the United States." *Geographical Review* 20 (1930): 207–223.

Gregg, Willis R. "Progress in Development of the U.S. Weather Service in Line with the Recommendations of the Science Advisory Board." *Science* 80 (1934): 349–351.

Gregg, Willis R. "Progress in Weather Forecasting." *Electrical Engineering* (October 1938): 405–412.

Gregg, Willis R. "Standard Atmosphere." NACA-TR-147. Washington, DC: National Advisory Committee for Aeronautics, 1922.

Gregg, Willis R. "The Weather Bureau and the Nation's Business." Radio talk typescript. Reichelderfer Records, NARA, Box 1.

Guerlac, Henry E. *RADAR in World War II*. Vol. 1, *Sections A–C*. New York: American Institute of Physics, 1987.

Guggenheim, Harry F. *Final Report of the Daniel Guggenheim Fund for the Promotion of Aeronautics, 1929*. New York: Guggenheim, 1930.

Guggenheim, Harry F. *Report of the Daniel Guggenheim Fund for the Promotion of Aeronautics, 1926 and 1927*. New York: Guggenheim, 1928.

Hales, Anton L. "Lloyd Viel Berkner (1905–1967)." In *Biographical Memoirs*, vol. 61, 3–26. Washington, DC: National Academy of Sciences, 1992.

Hallgren, Elisabeth Lynn. *The University Corporation for Atmospheric Research and the National Center for Atmospheric Research 1960–1970: An Institutional History.* Boulder, CO: University Corporation for Atmospheric Research, 1974.

Hallion, Richard R. *The Guggenheim Contribution to American Aviation.* Seattle: University of Washington Press, 1977.

Harper, Kristine C. *Weather by the Numbers: The Genesis of Modern Meteorology.* Cambridge, MA: MIT Press, 2008.

Harvard College Class of 1932. *Twenty-fifth Anniversary Report.* Cambridge, MA: Harvard University Press, 1957.

Haurwitz, Bernhard. "Geophysical Hydrodynamics." Lecture notes. Department of Atmospheric Science, Colorado State University, spring 1973.

Haurwitz, Bernhard. "Meteorology in the Twentieth Century: A Participant's View (Part II)." *Bulletin of the American Meteorological Society* 66 (4) (1985): 424–431.

Helland-Hansen, Bjørn, and Fridtjof Nansen. *The Norwegian Sea: Its Physical Oceanography Based on the Norwegian Researches, 1900–1904.* Report on Norwegian Fishery and Marine Investigations II, no. 2. Kristiania: Mallingske Bogtrykkeri, 1909.

Helmholtz, Hermann von. "On Atmospheric Movements" *Mechanics of the Earth's Atmosphere.* Translated by C. Abbe, 78–111. Washington, DC, 1891.

Helmholtz, Hermann von. "On the Integrals of the Hydro-dynamic Equations That Represent Vortex Motions, 1857." *Mechanics of the Earth's Atmosphere.* Translated by C. Abbe, 31–57. Washington, DC, 1891.

Helmholtz, Hermann von. "Wirbelstürme und Gewitter" (1875). *Vorträge und Reden.* Vol. 2, 139–164. Braunschweig: F. Vieweg, 1884.

Henry, Alfred J. *Weather Forecasting from Synoptic Charts.* Washington, DC: U.S. Department of Agriculture, 1930.

Hertz, Heinrich. *The Principles of Mechanics, Presented in a New Form.* Translated by D. E. Jones and J. T. Walley. London: Macmillan, 1899.

Hess, Victor F. "Unsolved Problems in Physics: Tasks for the Immediate Future in Cosmic Ray Studies." Nobel Lecture, December 12, 1936. http://www.nobelprize.org/nobel_prizes/physics/laureates/1936/hess-lecture.html.

Hinsdale, Guy. "Charles Frederick Marvin, 1858–1943." *Transactions of the American Clinical and Climatological Association* 58 (1946): liii–liv.

Holton, James R., Judith A. Curry , and J. A. Pyle, eds. 2003. *Encyclopedia of Atmospheric Sciences*, 6 vols. Amsterdam: Academic Press.

Hong, Sungook. *Wireless: From Marconi's Black-Box to the Audion.* Cambridge, MA: MIT Press, 2001.

Hubert, L. R., and Otto Berg. "A Rocket Portrait of a Tropical Storm." *Monthly Weather Review* 86 (1955): 119–124.

Hughes, Thomas Parke. *Networks of Power: Electrification in Western Society, 1880–1930.* Baltimore, MD: Johns Hopkins University Press, 1983.

Humphreys, William Jackson. "Willis Ray Gregg." *Science* n.s. 88 (1938): 318–319.

Jaw, Jeou-Jang. "The Formation of the Semipermanent Centers of Action in Relation to the Horizontal Solenoidal Field." *Journal of Meteorology* 3 (1946): 103–114.

Jeon, Chihyung. "Flying Weather Men and Robot Observers: Instruments, Inscriptions, and Identities in U.S. Upper-Air Observation, 1920–1940." *History and Technology* 26 (2) (2010): 119–145.

Khromov, Sergei Petrovich. *Einführung in die synoptische Wetteranalyse,* edited and translated by G. Swoboda. Vienna: J. Springer, 1940.

Koenigsberger, Leo. *Hermann von Helmholtz.* Translated by F. A. Welby. Oxford: Clarendon Press, 1906.

Kutzbach, Gisela. "Rossby, Carl-Gustaf Arvid." In *Complete Dictionary of Scientific Biography* vol. 11: 557–559. Detroit: Charles Scribner's Sons, 2008.

Kutzbach, Gisela. *The Thermal Theory of Cyclones: A History of Meteorological Thought in the Nineteenth Century.* Boston: American Meteorological Society, 1979.

Kvinge, Tor. *The "Conrad Holmboe" Expedition to East Greenland Waters in 1923.* Bergen: Norwegian Universities Press, 1963.

Lange, Karl O. "Radiometeorographs." *Bulletin of the American Meteorological Society* 16 (1935): 233–236, 267–271, 297–300, and 17 (1936): 136–148.

Laplace, Pierre-Simon Marquis de. "A Philosophical Essay on Probabilities." *Works* 7 (1): 6–7. Paris: Gauthier Villars, 1886.

Launius, R. D., J. R. Fleming, and D. H. DeVorkin, eds. *Globalizing Polar Science: Reconsidering the International Polar and Geophysical Years.* New York: Palgrave, 2010.

Lavine, Matthew. *The First Atomic Age: Scientists, Radiations, and the American Public, 1895–1945.* New York: Palgrave, 2013.

Leslie, Stuart W. "'A Different Kind of Beauty': Scientific and Architectural Style in I. M. Pei's Mesa Laboratory and Louis Kahn's Salk Institute." *Historical Studies in the Natural Sciences* 38 (2008): 173–221.

Lewis, John M. "Cal Tech's Program in Meteorology: 1933–1948." *Bulletin of the American Meteorological Society* 75 (1994): 69–81.

Lewis, John M. "Carl-Gustaf Rossby: A Study in Mentorship." *Bulletin of the American Meteorological Society* 73 (1992): 1425–1438.

Lewis, John M. "Clarifying the Dynamics of the General Circulation: Phillips's 1956 Experiment." *Bulletin of the American Meteorological Society* 79 (1998): 39–60.

Lewis, John M., Matthew G. Fearon, and Harold E. Klieforth. "Herbert Riehl: Intrepid and Enigmatic Scholar." *Bulletin of the American Meteorological Society* 93 (2012): 963–985.

Ligda, Myron G. H. "Radar Storm Observation." In *Compendium of Meteorology*, edited by T. F. Malone, 1265–1281. Boston: American Meteorological Society, 1951.

"Lindbergh's Mexican Trip Deferred by Bad Weather." *Washington Post*, December 12, 1927, 1.

Lindzen, R. S., E. N. Lorenz, and G. W. Platzman, eds. *The Atmosphere: A Challenge. The Science of Jule Gregory Charney*. Boston: American Meteorological Society, 1990.

Logsdon, John, ed. *Exploring the Unknown: Selected Documents in the History of the U.S. Civilian Space Program*. Vol. 2, *External Relationships*. Washington, DC: NASA History Office, 1996.

Logsdon, John, ed. "Meteorological Satellites." In *Exploring the Unknown: Selected Documents in the History of the U.S. Civil Space Program*. Vol. 3, *Using Space*, 156–158. Washington, DC: NASA History Office, 1998.

London, Julius. "Bernhard Haurwitz (1905–1986)." In *Biographical Memoirs*, vol. 69, 85–113. Washington, DC: National Academy of Sciences, 1996.

Lorenz, Edward N. "Deterministic Nonperiodic Flow." *Journal of the Atmospheric Sciences* 20 (1963): 130–141.

Lorenz, Edward N. *The Essence of Chaos*. Seattle: University of Washington Press, 1993.

Lorenz, Edward N. "The Evolution of Dynamic Meteorology." In *Historical Essays on Meteorology, 1919–1995*, edited by J. R. Fleming, 3–19. Boston: American Meteorological Society, 1996.

Lorenz, Edward N. "A History of Prevailing Ideas about the General Circulation of the Atmosphere." *Bulletin of the American Meteorological Society* 64 (1983): 730–734.

Lorenz, Edward N. "The Statistical Prediction of Solutions of Dynamic Equations." In *Proceedings of the International Symposium on Numerical Weather Prediction in Tokyo, November 7–13, 1960*, edited by S. Syono, 629–635. Tokyo: Meteorological Society of Japan, 1962.

Lynch, Peter. *The Emergence of Numerical Weather Prediction: Richardson's Dream*. Cambridge: Cambridge University Press, 2006.

Machta, Lester. "Finding the Site of the First Soviet Nuclear Test in 1949." *Bulletin of the American Meteorological Society* 73 (1992): 1797–1806.

Machta, Lester. "Nuclear Explosions and Meteorology." In *Proceedings of the Conference on Scientific Applications of Nuclear Explosions Held July 6–8, 1959, Los Alamos, New Mexico,* LAMS-2443, edited by G. A. Cowan, 26–30. Los Alamos, NM: Los Alamos Scientific Laboratory, 1960.

Machta, Lester. "Oral History Interview, with Spencer Weart and William Elliott." American Institute of Physics, OH 31417, April 25, 1991.

Machta, Lester, and D. L. Harris. "Effects of Atomic Explosions on Weather." *Science* n.s. 121 (1955): 75–81.

Malone, Thomas F. "International Cooperation in Meteorology and the Atmospheric Sciences." In *Research in Geophysics.* Vol. 1, *Sun, Upper Atmosphere, and Space,* edited by Hugh Odishaw, 533–540. Cambridge, MA: MIT Press, 1964.

Malone, Thomas F. "A National Institute for Atmospheric Research." *Eos, Transactions, American Geophysical Union* 40 (2) (1959): 95–111.

Manabe, Syukuro, and F. Möller. "On the Radiative Equilibrium and Balance of the Atmosphere." *Monthly Weather Review* 89 (1961): 503–532.

"Man's Milieu." *Time* 68, December 17, 1956, 68–79.

Marvin, Charles F. "Northern Hemisphere Map Interrupted." *Monthly Weather Review* 42 (7) (1914): 457.

Marvin, Charles F. "Status and Problems of Meteorology." *Proceedings of the National Academy of Sciences of the United States of America* 6 (1920): 561–572.

Massachusetts Institute of Technology. "A Brief History of the Department of Meteorology." *Bulletin of the American Meteorological Society* 32 (1951): 103–104.

Maynard, R. H. "Radar and Weather." *Journal of Meteorology* 2 (1945): 214–226.

Mazuzan, George T. "Up, Up, and Away: The Reinvigoration of Meteorology in the United States, 1958–1962." *Bulletin of the American Meteorological Society* 69 (1988): 1152–1163.

"Meeting of the International Commission for the Scientific Investigation of the Upper Air, at Bergen." *Meteorological Magazine* (September 1921): 215–217.

Meyer, William B. *Americans and Their Weather.* New York: Oxford University Press, 2000.

Miller, Howard S. "Science and Private Agencies." In *Science and Society in the United States,* edited by D. D. Van Tassel and M. G. Hall, 191–221. Homewood, IL: Dorsey, 1966.

Millikan, Robert A., "Preliminary Report of the Special Committee on the Weather Bureau of the Science Advisory Board." In "Work of the Weather Bureau," *Science* n.s. 78 (December 22, 1933): 582–585 and (December 29, 1933): 604–607.

Mills, Eric L. *The Fluid Envelope of Our Planet: How the Study of Ocean Currents Became a Science.* Toronto: University of Toronto Press, 2009.

Mindling, George W. "The Raymete and the Future." March 29, 1939. http://www.history.noaa.gov/art/weatherpoems1.html.

Mitra, S. K. "General Aspects of Upper Atmospheric Physics." In *Compendium of Meteorology*, edited by T. F. Malone. 245–261. Boston: American Meteorological Society, 1951.

Multhauf, Robert P. "Review of *A History of the United States Weather Bureau* by Donald R. Whitnah," *Technology and Culture* 3 (1962): 332–333.

Nebeker, Frederik. *Calculating the Weather: Meteorology in the Twentieth Century.* San Diego: Academic Press, 1995.

Needell, Alan A. *Science, Cold War and the American State: Lloyd Berkner and the Balance of Professional Ideals.* Amsterdam: Harwood Academic, 2000.

Neumann, Georg Hugo. "The Fourth Annual Conference in Atmospheric Chemistry, May 20–22, 1957." *Tellus* 10 (1958): 165–169.

"New Daily Weather Map." *Monthly Weather Review* 42 (1) (1914): 35–36.

New York University. "The Department of Meteorology, College of Engineering." *Bulletin of the American Meteorological Society* 32 (1951): 107–108.

Newton, Chester. "Roundtable Discussion about C. G. Rossby and the Chicago School, Jan. 31, 1996." American Meteorological Society Oral History Project, Archives, National Center for Atmospheric Research, Boulder, CO, 1996.

The Oak Leaf. Santa Rosa Junior College, Santa Rosa, CA, September 26, 1941.

O'Keefe, J. A., A. Eckels, and R. K. Squires. "Pear-Shaped Component of the Geoid from the Motion of *Vanguard 1.*" *Annals of the International Geophysical Year* 12 (1) (1960): 199–201.

Patrin. Santa Rosa Junior College, Santa Rosa, CA, 1930, 1938.

Persson, Anders. "Early Operational Numerical Weather Prediction outside the USA: An Historical Introduction. Part I: Internationalism and Engineering NWP in Sweden, 1952–69." *Meteorological Applications* 12 (2005): 135–159.

Persson, Anders. "Early Operational Numerical Weather Prediction outside the USA: An Historical Introduction. Part II: Twenty Countries around the World." *Meteorological Applications* 12 (2005): 269–289.

Petterssen, Sverre. "The Atmospheric Sciences 1961–1971." *Weatherwise* 185–187 (October 1962): 213.

Petterssen, Sverre. *Introduction to Meteorology.* New York: McGraw-Hill, 1941.

Petterssen, Sverre. *Weather Analysis and Forecasting: A Textbook on Synoptic Meteorology.* New York: McGraw-Hill, 1940.

Petterssen, Sverre. *Weathering the Storm: Sverre Petterssen, the D-Day Forecast and the Rise of Modern Meteorology.* Edited by J. R. Fleming. Boston: American Meteorological Society, 2001.

Pfeffer, Richard L., ed. *Dynamics of Climate: Proceedings of a Conference on the Application of Numerical Integration Techniques to the Problem of the General Circulation, October 26–28, 1955.* New York: Pergamon, 1960.

Phillips, Norman A. "Carl-Gustaf Rossby: His Times, Personality, and Actions." *Bulletin of the American Meteorological Society* 79 (1998): 1097–1112.

Phillips, Norman A. "The General Circulation of the Atmosphere: A Numerical Experiment." *Quarterly Journal of the Royal Meteorological Society* 82 (1956): 123–164.

Phillips, Norman A. "Interview by T. Hollingsworth, W. Washington, J. Tribbia, and A. Kasahara." American Meteorological Society Oral History Project, Archives, National Center for Atmospheric Research, 1989.

Pihl, Mogens. "Bjerknes, Carl Anton." *Complete Dictionary of Scientific Biography,* vol. 2: 166–167. Detroit: Charles Scribner's Sons, 2008.

Platzman, George W. *Conversations with Jule Charney.* NCAR Technical Note NCAR/ TN-298+PROC. Boulder, CO: National Center for Atmospheric Science; Cambridge, MA: Institute Archives and Special Collections, MIT Libraries, 1987.

Platzman, George W. "The ENIAC Computations of 1950: Gateway to Numerical Weather Prediction." *Bulletin of the American Meteorological Society* 60 (1979): 302–312.

Platzman, George W. "Some Remarks on High-Speed Automatic Computers and Their Use in Meteorology." *Tellus* 4 (1952): 168–178.

Poincaré, Jules Henri. *Les méthodes nouvelles de la mécanique céleste.* 3 vols. Paris: Gauthier-Villars, 1892–1899.

President's Science Advisory Committee. *Introduction to Outer Space.* Washington, DC: USGPO, 1958.

Purdom, James F. W., and W. Paul Menzel. "Evolution of Satellite Observations in the United States and Their Use in Meteorology." In *Historical Essays on Meteorology, 1919–1995,* edited by J. R. Fleming, 99–155. Boston: American Meteorological Society, 1996.

Reichelderfer, F. W. "The Atmosphere and the Sea in Motion." *WMO Bulletin* (July 1960): 139–144.

Reichelderfer, F. W. *Norwegian Methods of Weather Analysis.* Washington, DC: U.S. Department of Agriculture, 1932, 1934.

Reichhardt, Tony. "The First Photo from Space." *Air and Space Magazine* (November 2006).

Richardson, Lewis Fry. *Weather Prediction by Numerical Process.* Cambridge: Cambridge University Press, 1922; 2nd ed. 2007.

Riehl, Herbert. *Tropical Meteorology.* New York: McGraw-Hill, 1954.

Rogers, R. R., and P. L. Smith. "A Short History of Radar Meteorology." In *Historical Essays on Meteorology, 1919–1995,* edited by J. R. Fleming, 57–98. Boston: American Meteorological Society, 1996.

Rossby, C.-G. "Biographical Sketch." Typescript. MIT Museum, Cambridge, MA, 1939.

Rossby, C.-G. "Comments on Meteorological Research." *Journal of the Aeronautical Sciences* 1 (1934): 32–34.

Rossby, C.-G. "Current Problems in Meteorology." In *The Atmosphere and the Sea in Motion: Scientific Contributions to the Rossby Memorial Volume,* edited by B. Bolin, 9–50. New York: Rockefeller Institute Press and Oxford University Press, 1959.

Rossby, C.-G. "Den nordiska aerologiens arbetsuppgifter. En återblick och ett program." *Ymer* 43 (1923): 364–372.

Rossby, C.-G. "First Annual Report of the Meteorological Course of the Massachusetts Institute of Technology." Massachusetts Institute of Technology, Cambridge, MA, 1929.

Rossby, C.-G. "Horizontal Motion in the Atmosphere." *Journal of Meteorology* 1 (1944): 109–114.

Rossby, C.-G. "Meteorologiska Resultat av en Sommarseglats Runt de Brittiska Oarna." *Meddelanden Fran Statens Meteorologisk-Hydrografiska Anstalt* 3 (1). Stockholm: Norsted, 1925.

Rossby, C.-G. "Note on Activity during the Academic Years 1951–52 and 1952–1953 [at the Institute of Meteorology, University of Stockholm]." *Tellus* 5 (1953): 420–422.

Rossby, C.-G. "Note on Cooperative Research Projects." *Tellus* 3 (1951): 212–216.

Rossby, C.-G. "On the Origin of Travelling Discontinuities in the Atmosphere." *Geografiska Annaler* 6 (1924): 180–189.

Rossby, C.-G. "On the Propagation of Frequencies and Energy in Certain Types of Oceanic and Atmospheric Waves." *Journal of Meteorology* 2 (1945): 187–204.

Rossby, C.-G. "On the Solution of Problems of Atmospheric Motion by Means of Model Experiments." *Monthly Weather Review* 54 (1926): 237–240.

Rossby, C.-G. "Outline of a Research Program for the Meteorology Department, Massachusetts Institute of Technology." President's Report. Massachusetts Institute of Technology, Cambridge, MA, 1928.

Rossby, C.-G. "Planetary Flow Patterns in the Atmosphere." *Quarterly Journal of the Royal Meteorological Society* 66 (suppl.) (1940): 68–87.

Rossby, C.-G. "The Scientific Basis of Modern Meteorology." *Climate and Man: Yearbook of Agriculture*, 579–599. Washington, DC: U.S. Department of Agriculture, 1941. Reprinted in *Handbook of Meteorology*, edited by F. A. Berry, E. Bollay, and N. R. Beers, 501–529. New York: McGraw-Hill, 1945.

Rossby, C.-G. "Specific Evaluation of Pilot Balloon Data in Single Station Forecasting." Miscellaneous Reports No. 7. Institute of Meteorology, University of Chicago, 1943.

Rossby, C.-G. "Theory of Atmospheric Turbulence: A Historical Resume and an Outlook." *Monthly Weather Review* 55 (1927): 6–10.

Rossby, C.-G. and collaborators. "Relation between Variations in the Intensity of the Zonal Circulation of the Atmosphere and the Displacement of the Semipermanent Centers of Action." *Journal of Marine Research* 2 (1939): 38–55.

Rossby, C.-G., and H. Egner. "On the Chemical Climate and Its Variation with the Atmospheric Circulation Pattern." *Tellus* 7 (1955): 118–133.

Rossby, C.-G., V. J. Oliver, and M. Boyden. "Weather Estimates from Local Aerological Data: A Preliminary Report." Miscellaneous Reports No. 2. Institute of Meteorology, University of Chicago, 1942.

Rossby, C.-G., and Staff of the Department of Meteorology, University of Chicago. "On the General Circulation of the Atmosphere in the Middle Latitudes." *Bulletin of the American Meteorological Society* 28 (1947): 255–280.

Rossby, C.-G., and Richard Hanson Weightman, "Application of the Polar-Front Theory to a Series of American Weather Maps." *Monthly Weather Review* 54 (12) (December 1926): 485–496.

Rothschild, Rachel. "Burning Rain: The Long-Range Transboundary Air Pollution Project." In *Toxic Airs: Body, Place, Planet in Historical Perspective*, edited by J. R. Fleming and A. Johnson, 181–207. Pittsburgh: University of Pittsburgh Press, 2014.

Roulstone, Ian, and John Norbury. *Invisible in the Storm: The Role of Mathematics in Understanding Weather.* Princeton, NJ: Princeton University Press, 2013.

Rowland, Alex. "Introduction." In *A Guide to the Microfilm Edition of the Papers of the President's Science Advisory Committee, 1957–1961,* edited by Robert E. Lester. Frederick, MD: University Publications of America, 1991.

Sartz, R. S. N. "Norway's Contributions to the Natural Sciences." *Monthly Weather Review* 33 (1905): 539–540.

Sawyer, J. S. "Review of *Dynamic Meteorology and Weather Forecasting.*" *Quarterly Journal of the Royal Meteorological Society* 83 (1957): 560.

Schrenk, H. H., H. Heimann, G. D. Clayton, W. M. Gafafer, and Harry Wexler. "Air Pollution in Donora, Pa.: Epidemiology of the Unusual Smog Episode of October, 1948." *Public Health Bulletin* 306. Washington, DC: U.S. Public Health Service, 1949.

Schwerdtfeger, Werner. "Comments on Tor Bergeron's Contributions to Synoptic Meteorology." *Pageoph* 119 (1980–1981): 501–502.

Seaton, S. L. "The Ionosphere." In *Compendium of Meteorology,* edited by T. F. Malone, 334–340. Boston: American Meteorological Society, 1951.

Shapiro, M., and S. Grønås, eds. *The Life Cycle of Extratropical Cyclones.* Boston: American Meteorological Society, 1999.

Shaw, Sir Napier. *The Drama of Weather.* Cambridge: Cambridge University Press, 1933.

Sheppard, P. A. "Meteorology: The Exploration of the Upper Atmosphere." *Science Progress* 37 (1949): 488–503.

Simpson, George C. "Weather Forecasting." *Nature* 135 (1935): 703–705.

Smagorinsky, Joseph. "The Beginnings of Numerical Weather Prediction and General Circulation Modeling: Early Recollections." *Advances in Geophysics* 25 (1983): 3–37.

Smagorinsky, Margaret. "Interview, January 2, 2006, by Kristine Harper, Ronald Doel, Terry Smagorinsky Thompson." American Meteorological Society Oral History Project, Archives, National Center for Atmospheric Research, 2006.

Smith, Henry Ladd. *Airways: The History of Commercial Aviation in the United States.* New York: Knopf, 1942.

Snyder, Lynne Page. "The Death-Dealing Smog over Donora, Pennsylvania: Industrial Air Pollution, Public Health Policy, and the Politics of Expertise, 1948–1949." *Environmental History Review* 18 (1994): 117–139.

Spiegler, David B. "A History of Private Sector Meteorology." In *Historical Essays on Meteorology, 1919–1995,* edited by J. R. Fleming, 417–441. Boston: American Meteorological Society, 1996.

Stratton, Julius A. "The Effect of Rain and Fog on the Propagation of Very Short Radio Waves." *Proceedings of the Institute of Radio Engineers* 18 (1930): 1064–1074.

Sverdrup, Harald U. *Oceanography for Meteorologists.* New York: Prentice-Hall, 1942.

Sverdrup, Harald U., Martin W. Johnson, and Richard H. Fleming. *The Oceans: Their Physics, Chemistry, and General Biology.* New York: Prentice-Hall, 1942.

"The Telephone at the Paris Opera." *Scientific American,* December 31, 1881, 422–423.

Thompson, Philip D. "A History of Numerical Weather Prediction in the United States." *Bulletin of the American Meteorological Society* 64 (1983): 755–769.

Thompson, Philip D. "Interview by William Aspray." December 5, 1986, OH 125. Minneapolis: Charles Babbage Institute, 1986.

Trefil, James S., and Margaret Hindle Hazen. *Good Seeing: A Century of Science at the Carnegie Institution of Washington.* Washington, DC: Joseph Henry Press, 2002.

"Twelve Colleges Plan Weather Studies." *New York Times,* October 10, 1958, 33.

Uda, Michitako, Hiroshima District Central Meteorological Observatory. "Meteorological Conditions Related to the Atomic Bomb Explosion at the Hiroshima City." Original in Japanese. Atomic Bomb Casualty Reports No. 155, Record Group 331, Military Agency Records, box 7409, Supreme Commander for the Allied Powers (SCAP). Economic and Scientific Section. Scientific and Technical Division, U.S. National Archives and Records Administration, November 1947.

University Committee on Atmospheric Research. *Preliminary Plans for a National Institute for Atmospheric Research.* Washington, DC: National Science Foundation, 1959.

University Corporation for Atmospheric Research (UCAR). *Remembering Walt Roberts.* Boulder, CO: UCAR, 1991.

University Corporation for Atmospheric Research (UCAR). *UCAR at Twenty-five.* Boulder, CO: UCAR, 1985. http://www.ucar.edu/communications/ucar25/index .html.

United Nations Resolution 1721 (XVI), International Co-operation in the Peaceful Uses of Outer Space, 1961, https://en.wikisource.org/wiki/United_Nations_General _Assembly_Resolution_1962.

United States National Academy of Sciences, Committee on Atmospheric Sciences. *The Atmospheric Sciences, 1961–1971.* A Report to the Special Assistant to the President for Science and Technology. Vol. 1, *Goals and Plans;* Vol. 2, *Summaries of Planning Conferences;* Vol. 3, *Goals and Plans for Aeronomy.* Washington, DC: National Academy of Sciences, 1962.

United States National Academy of Sciences, Committee on Atmospheric Sciences. *Meteorology on the Move: A Progress Report*. Washington, DC: National Academy of Sciences, 1960.

United States National Academy of Sciences, Committee on Atmospheric Sciences. *Proceedings of the Scientific Information Meeting on Atmospheric Sciences, November 24, 1958*. National Academy of Sciences, Washington, DC, 1959.

United States National Academy of Sciences, Committee on Meteorology. *Research and Education in Meteorology: An Interim Report*. Washington, DC: National Academy of Sciences–National Research Council, 1958.

United States National Science Foundation. *Weather and Climate Modification*. NSF 66–3. Washington, DC: National Science Foundation, 1965.

United States Senate. *Organizing for National Security, Science Organization and the President's Office: A Study Submitted to the Committee on Government Operation*. Washington: USGPO, 1961.

United States Weather Bureau. "Report on Alert Number 112 of the Atomic Detection System, September 29, 1949," marked Top Secret. Digital National Security Archive.

United States Weather Bureau. *Reports of the Chief*. Washington, DC: USGPO, 1913–1934.

Van Allen, J. A., and L. A. Frank. "Radiation around the Earth to a Radial Distance of 107,400 km." *Nature* 183 (1959): 430–434.

Van Allen, J. A., G. H. Ludwig, E. C. Ray, and C. E. McIlwain. "Observation of High Intensity Radiation by Satellites 1958 Alpha and Gamma." *U.S. National Academy of Sciences, I.G.Y. Satellite Report Series* 3 (1958): 73–92.

Van Bebber, W. J. *Lehrbuch der Meteorologie für Studierende und zum Gebrauche in der Praxis*. Stuttgart: F. Enke, 1890.

Vervarslinga på Vestlandet 25 år Festskrift. Bergen: A. S. John Griegs Boktrykkeri, 1944.

Von Neumann, John. "Can We Survive Technology?" *Fortune* (June 1955): 106–108.

Walker, Gabrielle. *An Ocean of Air: A Natural History of the Atmosphere*. London: Bloomsbury, 2007.

"Walter Orr Roberts Is Dead at Seventy-four: Expert on Climate's Effect on Life." *New York Times*, March 14, 1990.

Weightman, R. H. *Forecasting from Synoptic Weather Charts*. Washington, DC: U.S. Department of Agriculture, 1936.

Weightman, R. H., and C.-G. Rossby. "Polar Fronts in the United States." *Bulletin of the American Meteorological Society* 8 (1927): 32–33.

Wenstrom, William H. "Radiometeorography as Applied to Unmanned Balloons." *Monthly Weather Review* 62 (1934): 221–226.

Wexler, Harry. "Analysis of a Warm-Front-Type Occlusion." *Monthly Weather Review* 63 (1935): 213–221.

Wexler, Harry. "Comments on Meteorological Control of Air Pollution." In *Air Pollution: Proceedings of the United States Technical Conference on Air Pollution*, 833–834. New York: McGraw-Hill, 1952.

Wexler, Harry. "A Comparison of the Linke and Ångstrom Measures of Atmospheric Turbidity and Their Application to North American Air Masses." *Eos, Transactions, American Geophysical Union* 14 (April 1933): 92–99.

Wexler, Harry. "Cooling in the Lower Atmosphere and the Structure of Polar Continental Air." *Monthly Weather Review* 64 (1936): 122–136.

Wexler, Harry. "Deflections Produced in a Tropical Current by Its Flow over a Polar Wedge." *Bulletin of the American Meteorological Society* 16 (1935): 250–252.

Wexler, Harry. "Formation of Polar Anticyclones." *Monthly Weather Review* 65 (1937): 229–236.

Wexler, Harry. "Global Meteorology and the United Nations." *Weatherwise* 15 (4) (August 1962): 141–147.

Wexler, Harry. "The Great Smoke Pall: September 24–30, 1950." *Weatherwise* 3 (6) (December 1950): 129–134, 142.

Wexler, Harry. "Meet the Division Chiefs: Scientific Services." *Weather Bureau Topics* 10 (12) (1951): 226–228.

Wexler, Harry. "Meteorological Aspects of Atomic Radiation." *Science* 124 (1956): 1050–1112.

Wexler, Harry. ""Meteorology and Air Pollution." *American Journal of Public Health*, Part 2, *Yearbook* 41 (5) (May 1951): 88–91.

Wexler, Harry. "Meteorology in the International Geophysical Year." *Scientific Monthly* 84 (1957): 141–145.

Wexler, Harry. "Modifying Weather on a Large Scale." *Science*, n.s., 128 (1958): 1059–1063.

Wexler, Harry. "Observations of Nocturnal Radiation at Fairbanks, Alaska, and Fargo, N. Dak." *Monthly Weather Review Supplement* 46 (1941).

Wexler, Harry. "Observed Transverse Circulations in the Atmosphere and their Climatological Implications." ScD thesis, Department of Aeronautical Science, Massachusetts Institute of Technology, June 1939.

Wexler, Harry. "Observing the Weather from a Satellite Vehicle." *JBIS, Journal of the British Interplanetary Society* 13 (1954): 269–276.

Wexler, Harry. "Preface." *Meteorology and Atomic Energy*. Prepared by the U.S. Weather Bureau for the U.S. Atomic Energy Commission, Washington, DC, 1955.

Wexler, Harry. "The Role of Meteorology in Air Pollution." *Air Pollution*, 49–61. New York: Columbia University Press, World Health Organization, 1961.

Wexler, Harry. "The Satellite and Meteorology." *Journal of Astronautics* 4 (Spring 1957): 1–6.

Wexler, Harry. "Structure of Hurricanes as Determined by Radar." *Annals of the New York Academy of Sciences* 48 (1947): 821–844.

Wexler, Harry. "The Structure of the September, 1944 Hurricane When off Cape Henry, Virginia." *Bulletin of the American Meteorological Society* 26 (May 1945): 156–159.

Wexler, Harry. "Turbidities of American Air Masses and Conclusions Regarding the Seasonal Variation in Atmospheric Dust Content." *Monthly Weather Review* 62 (1934): 397–402.

Wexler, Harry, Melvin Brodshaug, and Donald C. Doane. *Weather Elements (The Polar Front Theory): A Guide for Use with the Instructional Sound Film "The Weather."* Chicago: University of Chicago Press, 1942.

Wexler, Harry, and Bernhard Haurwitz. "Trübungsfaktoren Nordamerikanischer Luftmassen." *Meteorologische Zeitschrift* 51 (1934): 236–238.

Wexler, Harry, L. Machta, D. H. Pack, and F. D. White. "Atomic Energy and Meteorology." *International Conference on the Peaceful Uses of Atomic Energy*. A/CONF. 8/ P276. Geneva, 1955.

Wexler, Harry, and L. W. Sheridan. *A Study of Barograph Records for July 16, 1945, in the Vicinity of the New Mexico Atomic Bomb Explosion*. Washington, DC: U.S. Weather Bureau, 1946.

Wexler, Harry, collaborator. "The Weather." Film, 14 mins. Chicago: Encyclopedia Britannica Films, 1941.

Wexler, Raymond. "Radar Detection of a Frontal Storm 18 June 1946." *Journal of Meteorology* 4 (1947): 38–44.

Whitnah, Donald R. *A History of the United States Weather Bureau*. Urbana: University of Illinois Press, 1961.

Widger, William K., Jr. "Examples of Project *TIROS* Data and Their Practical Meteorological Use." *GRD Research Notes* 38 (July 1960).

Wilczek, Frank. "Beautiful Losers: Kelvin's Vortex Atoms." http://www.pbs.org/wgbh/nova/blogs/physics/2011/12/beautiful-losers-kelvins-vortex-atoms.

Willett, Hurd. "C.-G. Rossby, Leader of Modern Meteorology." *Science* 127 (3300) (1958): 686–687.

Willett, Hurd. "Dynamic Meteorology." *Physics of the Earth III: Meteorology*, 133–233. Washington, DC: National Research Council, 1931.

Willis, E. P., and W. H. Hooke. "Cleveland Abbe and American Meteorology, 1871–1901." *Bulletin of the American Meteorological Society* 87 (2006): 315–326.

Wood, F. B. "A Flight into the September, 1944 Hurricane off Cape Henry, Virginia." *Bulletin of the American Meteorological Society* 26 (1945): 153–156.

Yalda, Sepideh. "Wexler, Harry." *Complete Dictionary of Scientific Biography*, vol. 25, 273–276. Detroit: Charles Scribner's Sons, 2008.

Yaskell, Steven Haywood. "From Research Institution to Astronomical Museum: A History of the Stockholm Observatory." *Journal of Astronomical History and Heritage* 11, no 2 (2008):146–155.

Yoden, Shigeo. "Atmospheric Predictability." *Journal of the Meteorological Society of Japan* 85B (2007): 77–102.

Zworykin, V. K. "Outline of Weather Proposal, October 1945." Princeton, NJ: RCA Laboratories, 1945, copy in Wexler Papers, 18. Reproduced in *History of Meteorology* 4 (2008): 57–78.

Index